PIRATES ENHANCED CLASSICS

HOW TO STOP WORRYING AND START LIVING

DALE CARNEGIE

PIRATES
INDIA

HOW TO STOP WORRYING AND START LIVING

Published by Pirates

Office Address:
Sanghi Brothers
Station Road
Jodhpur 342001

Ordering Information:
Quantity sales: Special discounts are available on quantity purchases by corporations, associations, and others.
For details, contact the publisher at the Office Address above.

Orders by trade bookstores and wholesalers.
Please contact Pirates Distribution: Tel: (+91) 9999 871 832; or Email: sales@pirates.ind.in

Cover Design by Pirates

Printed and bound in India

3/23

For select territories only.

Contents:

PART EIGHT - HOW TO FIND THE KIND OF WORK IN WHICH YOU MAY BE HAPPY AND SUCCESSFUL

PART NINE - HOW TO LESSEN YOUR FINANCIAL WORRIES

PART TEN - "HOW I CONQUERED WORRY" (32 TRUE STORIES)

Sixteen Ways in Which This Book Will Help You

1. Gives you a number of practical, tested formulas for solving worry situations.
2. Shows you how to eliminate fifty per cent of your business worries immediately.
3. Brings you seven ways to cultivate a mental attitude that will bring you peace and happiness.
4. Shows you how to lessen financial worries.
5. Explains a law that will outlaw many of your worries.
6. Tells you how to turn criticism to your advantage.
7. Shows how the housewife can avoid fatigue-and keep looking young.
8. Gives four working habits that will help prevent fatigue and worry.
9. Tells you how to add one hour a day to your working life.
10. Shows you how to avoid emotional upsets.
11. Gives you the stories of scores of everyday men and women, who tell you in their own words how they stopped worrying and started living.
12. Gives you Alfred Adler's prescription for curing melancholia in fourteen days.
13. Gives you the 21 words that enabled the world-famous physician, Sir William Osier, to banish worry.
14. Explains the three magic steps that Willis H. Carrier, founder of the air-conditioning industry, uses to conquer worry.
15. Shows you how to use what William James called "the sovereign cure for worry".
16. Gives you details of how many famous men conquered worry-men like Arthur Hays Sulzberger, publisher of the New York Times; Herbert E. Hawkes, former Dean of Columbia University; Ordway Tead, Chairman of the Board of Higher Education, New York City; Jack Dempsey; Connie Mack; Roger W. Babson; Admiral Byrd; Henry Ford; Gene Autry; J.C. Penney; and John D. Rockefeller.

Preface

How This Book Was Written-and Why

Thirty-Five years ago, I was one of the unhappiest lads in New York. I was selling motor-trucks for a living. I didn't know what made a motor-truck run. That wasn't all: I didn't want to know. I despised my job. I despised living in a cheap furnished room on West Fifty-sixth Street-a room infested with cockroaches. I still remember that I had a bunch of neckties hanging on the walls; and when I reached out of a morning to get a fresh necktie, the cockroaches scattered in all directions. I despised having to eat in cheap, dirty restaurants that were also probably infested with cockroaches.

I came home to my lonely room each night with a sick headache-a headache bred and fed by disappointment, worry, bitterness, and rebellion. I was rebelling because the dreams I had nourished back in my college days had turned into nightmares. Was this life? Was this the vital adventure to which I had looked forward so eagerly? Was this all life would ever mean to me-working at a job I despised, living with cockroaches, eating vile food-and with no hope for the future? ... I longed for leisure to read, and to write the books I had dreamed of writing back in my college days.

I knew I had everything to gain and nothing to lose by giving up the job I despised. I wasn't interested in making a lot of money, but I was interested in making a lot of living. In short, I had come to the Rubicon-to that moment of decision which faces most young people when they start out in life. So I made my decision-and that decision completely altered my future. It has made the last thirty-five years happy and rewarding beyond my most Utopian aspirations.

My decision was this: I would give up the work I loathed; and, since I had spent four years studying in the State Teachers' College at Warrensburg, Missouri, preparing to teach, I would make my living teaching adult classes in night schools. Then I would have my days free to read books, prepare lectures, write novels and short stories. I wanted "to live to write and write to live".

What subject should I teach to adults at night? As I looked back and evaluated my own college training, I saw that the training and experience I had had in public speaking had been of more practical value to me in business-and in life-than everything else I had studied in college all put together. Why? Because it had wiped out my timidity and lack of confidence and given me the courage and assurance to deal with people. It had also made clear that leadership usually gravitates to the man who can get up and say what he thinks I applied for a position teaching public speaking in the night extension courses both at Columbia University and New York University, but these universities decided they could struggle along somehow without my help.

I was disappointed then-but I now thank God that they did turn me down, because I started teaching in Y.M.C.A. night schools, where I had to show concrete results and show them quickly. What a challenge that was! These adults didn't come to my classes because they wanted college credits or social prestige. They came for one reason only: they wanted to solve their problems. They wanted to be able to stand up on their own feet and say a few words at a business meeting without fainting from fright. Salesmen wanted to be able to call on a tough customer without having to walk around the block three times to get up courage. They wanted to develop poise and self-confidence. They wanted to get ahead in business. They wanted to have more money for their families. And since they were paying their tuition on an installment basis-and they stopped paying if they didn't get results-and since I was being paid, not a salary, but a percentage of the profits, I had to be practical if I wanted to eat.

I felt at the time that I was teaching under a handicap, but I realise now that I was getting priceless training. I had to motivate my students. I had to help them solve their problems.

I had to make each session so inspiring that they wanted to continue coming.

It was exciting work. I loved it. I was astounded at how quickly these businessmen developed selfconfidence and how quickly many of them secured promotions and increased pay. The classes were succeeding far beyond my most optimistic hopes. Within three seasons, the Y.M.C.A.s, which had refused to pay me five dollars a night in salary, were paying me thirty dollars a night on a percentage basis. At first, I taught only public speaking, but, as the years went by, I saw that these adults also needed the ability to win friends and influence people. Since I couldn't find an adequate textbook on human relations, I wrote one myself. It was written-no, it wasn't written in the usual way. It grew and evolved out of the experiences of the adults in these classes. I called it How to Win Friends and Influence People.

Since it was written solely as a textbook for my own adult classes, and since I had written four other books that no one had ever heard of, I never dreamed that it would have a large sale: I am probably one of the most astonished authors now living.

As the years went by, I realised that another one of the biggest problems of these adults was worry. A large majority of my students were businessmen-executives, salesmen, engineers, accountants: a cross section of all the trades and professions-and most of them had problems! There were women in the classes-business women and housewives. They, too, had problems! Clearly, what I needed was a textbook on how to conquer worry-so again I tried to find one. I went to New York's great public library at Fifth Avenue and Forty-second Street and discovered to my astonishment that this library had only twenty-two books listed under the title WORRY. I also noticed, to my amusement, that it had one hundred and eighty-nine books listed under WORMS. Almost nine times as many books about worms as about worry! Astounding, isn't it? Since worry is one of the biggest problems facing mankind, you would think, wouldn't you, that every high school and college in the land would give a course on "How to Stop Worrying"?

Yet, if there is even one course on that subject in any college in the land, I have never heard of it. No wonder David Seabury said in his book *How to Worry Successfully*: "We come to maturity with as little preparation for the pressures of experience as a bookworm asked to do a ballet."

The result? More than half of our hospital beds are occupied by people with nervous and emotional troubles.

I looked over those twenty-two books on worry reposing on the shelves of the New York Public Library. In addition, I purchased all the books on worry I could find; yet I couldn't discover even one that I could use as a text in my course for adults. So I resolved to write one myself.

I began preparing myself to write this book seven years ago. How? By reading what the philosophers of all ages have said about worry. I also read hundreds of biographies, all the way from Confucius to Churchill. I also interviewed scores of prominent people in many walks of life, such as Jack Dempsey, General Omar Bradley, General Mark Clark, Henry Ford, Eleanor Roosevelt, and Dorothy Dix. But that was only a beginning.

I also did something else that was far more important than the interviews and the reading. I worked for five years in a laboratory for conquering worry - a laboratory conducted in our own adult classes. As far as I know, it is the first and only laboratory of its kind in the world. This is what we did. We gave students a set of rules on how to stop worrying and asked them to apply these rules in their own lives and then talk to the class on the results they had obtained. Others reported on techniques they had used in the past.

As a result of this experience, I presume I have listened to more talks on "How I Conquered Worry" than has any other individual who ever walked this earth. In addition, I read hundreds of other talks on "How I Conquered Worry" talks that were sent to me by mail-talks that had won prizes in our classes that are held in more than a hundred and seventy cities throughout the United States and Canada. So this book didn't come out of an ivory tower. Neither is it an academic preachment on how worry might be conquered. Instead, I have tried to write a fast-moving, concise, documented report on how worry has been conquered by thousands of adults. One thing is certain: this book is practical. You can set your teeth in it.

I am happy to say that you won't find in this book stories about an imaginary "Mr. B--" or a vague "Mary and John" whom no one can identify. Except in a few rare cases, this book names names and gives street addresses. It is authentic. It is documented. It is vouched for-and certified.

"Science," said the French philosopher Valery, "is a collection of successful recipes." That is what this book is, a collection of successful and time-tested recipes to rid our lives of worry. However, let me warn you: you won't find anything new in it, but you will find much that is not generally applied. And when it comes to that, you and I don't need to be told anything new. We already know enough to lead perfect lives. We have all read the golden rule and the Sermon on the Mount. Our trouble is not ignorance, but inaction. The purpose of this book is to restate, illustrate, streamline, air-condition, and glorify a lot of ancient and basic truths-and kick you in the shins and make you do something about applying them.

You didn't pick up this book to read about how it was written. You are looking for action. All right, let's go. Please read the first forty-four pages of this book-and if by that time you don't feel that you have acquired a new power and a new inspiration to stop worry and enjoy life-then toss this book into the dust-bin. It is no good for you.

DALE CARNEGIE

Nine Suggestions on How to Get the Most Out of This Book

1 If you wish to get the most out of this book, there is one indispensable requirement, one essential infinitely more important than any rules or technique. Unless you have this one fundamental requisite a thousand rules on how to study will avail little. And if you do have this cardinal endowment, then you can achieve wonders without reading any suggestions for getting the most out of a book.

What is this magic requirement? Just this: a deep, driving desire to learn, a vigorous determination to stop worrying and start living.

How can you develop such an urge? By constantly reminding yourself of how important these principles are to you. Picture to yourself how their mastery will aid you in living a richer, happier life. Say to yourself over and over: "My peace of mind, my happiness, my health, and perhaps even my income will, in the long run, depend largely on applying the old, obvious, and eternal truths taught in this book."

2 Read each chapter rapidly at first to get a bird's-eye view of it. You will probably be tempted then to rush on to the next one. But don't. Unless you are reading merely for entertainment. But if you are reading because you want to stop worrying and start living, then go back and re-read each chapter thoroughly. In the long run, this will mean saving time and getting results.

3 Stop frequently in your reading to think over what you are reading. Ask yourself just how and when you can apply each suggestion. That kind of reading will aid you far more than racing ahead like a whippet chasing a rabbit.

4 Read with a red crayon, pencil, or fountain-pen in your hand; and when you come across a suggestion that you feel you can use, draw a line beside it. If it is a four-star suggestion, then underscore every sentence, or mark it with "XXXX". Marking and underscoring a book make it more interesting, and far easier to review rapidly.

5 I know a man who has been office manager for a large insurance concern for fifteen years. He reads every month all the insurance contracts his company issues. Yes, he reads the same contracts over month after month, year after year. Why? Because experience has taught him that that is the only way he can keep their provisions clearly in mind.

I once spent almost two years writing a book on public speaking; and yet I find I have to keep going back over it from time to time in order to remember what I wrote in my own book. The rapidity with which we forget is astonishing.

So, if you want to get a real, lasting benefit out of this book, don't imagine that skimming through it once will suffice. After reading it thoroughly, you ought to spend a few hours reviewing it every month. Keep it on your desk in front of you every day. Glance through it often. Keep constantly impressing yourself with the rich possibilities for improvement that still lie in the offing. Remember that the use of these principles can be made habitual and unconscious only by a constant and vigorous campaign of review and application. There is no other way.

6 Bernard Shaw once remarked: "If you teach a man anything, he will never learn." Shaw was right. Learning is an active process. We learn by doing. So, if you desire to master the principles you are studying in this book, do something about them. Apply these rules at every opportunity. If you don't you will forget them quickly. Only knowledge that is used sticks in your mind.

You will probably find it difficult to apply these suggestions all the time. I know, because I wrote this book, and yet frequently I find it difficult to apply everything I have advocated here. So, as you read this book, remember that you are not merely trying to acquire information. You are attempting to form new habits.

Ah yes, you are attempting a new way of life. That will require time and persistence and daily application.

So refer to these pages often. Regard this as a working handbook on conquering worry; and when you are confronted with some trying problem, don't get all stirred up. Don't do the natural thing, the impulsive thing. That is usually wrong.

Instead, turn to these pages and review the paragraphs you have underscored. Then try these new ways and watch, them achieve magic for you.

7 Offer your wife a shilling every time she catches you violating one of the principles advocated in this book. She will break you!

8 Please turn to pages 193-4 of this book and read how the Wall Street banker, H.P. Howell, and old Ben Franklin corrected their mistakes. Why don't you use the Howell and Franklin techniques to check up on your application of the principles discussed in this book? If you do, two things will result.

First, you will find yourself engaged in an educational process that is both intriguing and priceless.

Second, you will find that your ability to stop worrying and start living will grow and spread like a green bay tree.

9 Keep a diary-a diary in which you ought to record your triumphs in the application of these principles. Be specific. Give names, dates, results. Keeping such a record will inspire you to greater efforts; and how fascinating these entries will be when you chance upon them some evening, years from now!

IN A NUTSHELL

1 **GOLDEN RULE 1**

Develop a deep, driving desire to master the principles of conquering worry.

2 **GOLDEN RULE 2**

Read each chapter twice before going on to the next one.

3 **GOLDEN RULE 3**

As you read, stop frequently to ask yourself how you can apply each suggestion.

4 **GOLDEN RULE 4**

Underscore each important idea.

5 **GOLDEN RULE 5**

Review this book each month.

GOLDEN RULE 6

Apply these principles at every opportunity. Use this volume as a working handbook to help you solve your daily problems.

GOLDEN RULE 7

Make a lively game put of your learning by offering some friend a shilling every time he catches you violating one of these principles.

GOLDEN RULE 8

Check up each week on the progress you are making. Ask yourself what mistakes you have made, what improvement, what lessons you have learned for the future.

GOLDEN RULE 9

Keep notes in the back of this book showing how and when you have applied these principles.

PART ONE

Fundamental Facts You Should Know About Worry

Chapter 1

Live in "Day-tight Compartments"

In the spring of 1871, a young man picked up a book and read twenty-one words that had a profound effect on his future. A medical student at the Montreal General Hospital, he was worried about passing the final examination, worried about what to do, where to go, how to build up a practice, how to make a living.

The twenty-one words that this young medical student read in 1871 helped him to become the most famous physician of his generation. He organised the world-famous Johns Hopkins School of Medicine. He became Regius Professor of Medicine at Oxford-the highest honour that can be bestowed upon any medical man in the British Empire. He was knighted by the King of England. When he died, two huge volumes containing 1,466 pages were required to tell the story of his life.

His name was Sir William Osier. Here are the twenty-one words that he read in the spring of 1871- twenty-one words from Thomas Carlyle that helped him lead a life free from worry: "Our main business is not to see what lies dimly at a distance, but to do what lies clearly at hand."

Forty-two years later, on a soft spring night when the tulips were blooming on the campus, this man, Sir William Osier, addressed the students of Yale University. He told those Yale students that a man like himself who had been a professor in four universities and had written a popular book was supposed to have "brains of a special quality". He declared that that was untrue. He said that his intimate friends knew that his brains were "of the most mediocre character".

What, then, was the secret of his success? He stated that it was owing to what he called living in "daytight compartments." What did he mean by that? A few months before he spoke at

Yale, Sir William Osier had crossed the Atlantic on a great ocean liner where the captain standing on the bridge, could press a button and-presto!-there was a clanging of machinery and various parts of the ship were immediately shut off from one another-shut off into watertight compartments. "Now each one of you," Dr. Osier said to those Yale students, "is a much more marvelous organisation than the great liner, and bound on a longer voyage. What I urge is that you so learn to control the machinery as to live with 'day-tight compartments' as the most certain way to ensure safety on the voyage. Get on the bridge, and see that at least the great bulkheads are in working order. Touch a button and hear, at every level of your life, the iron doors shutting out the Past - the dead yesterdays. Touch another and shut off, with a metal curtain, the Future - the unborn tomorrows. Then you are safe - safe for today! ... Shut off the past! Let the dead past bury its dead. ... Shut out the yesterdays which have lighted fools the way to dusty death. ... The load of tomorrow, added to that of yesterday, carried today, makes the strongest falter. Shut off the future as tightly as the past. ... The future is today. ... There is no tomorrow. The day of man's salvation is now. Waste of energy, mental distress, nervous worries dog the steps of a man who is anxious about the future. ... Shut close, then the great fore and aft bulkheads, and prepare to cultivate the habit of life of 'day-tight compartments'."

Did Dr. Osier mean to say that we should not make any effort to prepare for tomorrow? No. Not at all. But he did go on in that address to say that the best possible way to prepare for tomorrow is to concentrate with all your intelligence, all your enthusiasm, on doing today's work superbly today. That is the only possible way you can prepare for the future.

Sir William Osier urged the students at Yale to begin the day with Christ's prayer: "Give us this day our daily bread."

Remember that that prayer asks only for today's bread. It doesn't complain about the stale bread we had to eat yesterday; and it doesn't say: "Oh, God, it has been pretty dry out in the wheat belt lately and we may have another drought-and then how will I get bread to eat next autumn-or suppose I lose my job-oh, God, how could I get bread then?"

No, this prayer teaches us to ask for today's bread only. Today's bread is the only kind of bread you can possibly eat.

Years ago, a penniless philosopher was wandering through a stony country where the people had a hard time making a living. One day a crowd gathered about him on a hill, and he gave what is probably the most-quoted speech ever delivered anywhere at any time. This speech contains twentysix words that have gone ringing down across the centuries: "Take therefore no thought for the morrow; for the morrow shall take thought for the things of itself. Sufficient unto the day is the evil thereof."

Many men have rejected those words of Jesus: "Take no thought for the morrow." They have rejected those words as a counsel of perfection, as a bit of Oriental mysticism. "I must take thought for the morrow," they say. "I must take out insurance to protect my family. I must lay aside money for my old age. I must plan and prepare to get ahead."

Right! Of course you must. The truth is that those words of Jesus, translated over three hundred years ago, don't mean today what they meant during the reign of King James. Three hundred years ago the word thought frequently meant anxiety. Modern versions of the Bible quote Jesus more accurately as saying: "Have no anxiety for the tomorrow."

By all means take thought for the tomorrow, yes, careful thought and planning and preparation. But have no anxiety.

During the war, our military leaders planned for the morrow, but they could not afford to have any anxiety. "I have supplied the best men with the best equipment we have," said Admiral Ernest J. King, who directed the United States Navy, "and have given them what seems to be the wisest mission. That is all I can do."

"If a ship has been sunk," Admiral King went on, "I can't bring it up. If it is going to be sunk, I can't stop it. I can use my time much better working on tomorrow's problem than by fretting about yesterday's. Besides, if I let those things get me, I wouldn't last long."

Whether in war or peace, the chief difference between good thinking and bad thinking is this: good thinking deals with causes and effects and leads to logical, constructive planning; bad thinking frequently leads to tension and nervous breakdowns.

I recently had the privilege of interviewing Arthur Hays Sulzberger, publisher of one of the most famous newspapers in the world, *The New York Times*. Mr. Sulzberger told me that when the Second World War flamed across Europe, he was so stunned, so worried about the future, that he found it almost impossible to sleep. He would frequently get out of bed in the middle of the night, take some canvas and tubes of paint, look in the mirror, and try to paint a portrait of himself. He didn't know anything about painting, but he painted anyway, to get his mind off his worries. Mr. Sulzberger told me that he was never able to banish his worries and find peace until he had adopted as his motto five words from a church hymn: One step enough for me.

Lead, kindly Light ...
Keep thou my feet: I do not ask to see
The distant scene; one step enough for me.

At about the same time, a young man in uniform-somewhere in Europe-was learning the same lesson. His name was Ted Bengermino, of 5716 Newholme Road, Baltimore, Maryland-and he had worried himself into a first-class case of combat fatigue.

"In April, 1945," writes Ted Bengermino, "I had worried until I had developed what doctors call a 'spasmodic transverse colon'-a condition that produced intense pain. If the war hadn't ended when it did, I am sure I would have had a complete physical breakdown.

"I was utterly exhausted. I was a Graves Registration, Non-commissioned Officer for the 94th Infantry Division. My work was to help set up and maintain records of all men killed in action, missing in action, and hospitalised. I also had to help disinter the bodies of both Allied and enemy soldiers who had been killed and hastily buried in shallow graves during the pitch of battle. I had to gather up the personal effects of these men and see that they were sent back to parents or closest relatives who would prize these personal effects so much. I was constantly worried for fear we might be making embarrassing and serious mistakes. I was worried about whether or not I would come through all this. I was worried about whether I would live to hold my only child in my arms-a son of sixteen months, whom I had never seen. I was so worried and exhausted that I lost thirty-four pounds. I was so frantic that I was almost out of my mind. I looked at my hands. They were hardly more than skin and bones. I was terrified at the thought of going home a physical wreck. I broke down and sobbed like a child. I was so shaken that tears welled up every time I was alone. There was one period soon after the Battle of the Bulge started that I wept so often that I almost gave up hope of ever being a normal human being again.

"I ended up in an Army dispensary. An Army doctor gave me some advice which has completely changed my life. After giving me a thorough physical examination, he informed me that my troubles were mental. 'Ted,' he said, 'I want you to think of your life as an hourglass. You know there are thousands of grains of sand in the top of the hourglass; and they all pass slowly and evenly through the narrow neck in the middle. Nothing you or I could do would make more than one grain of sand pass through this narrow neck without impairing the hourglass. You and I and everyone else are like this hourglass. When we start in the morning, there are hundreds of tasks which we feel that we must accomplish that day, but if we do not take them one at a time and let them pass through the day slowly and evenly, as do the grains of sand passing through the narrow neck of the hourglass, then we are bound to break our own physical or mental structure.'

"I have practised that philosophy ever since that memorable day that an Army doctor gave it to me. 'One grain of sand at a time. ... One task at a time.' That advice saved me physically and mentally during the war; and it has also helped me in my present position in business. I am a Stock Control Clerk for the Commercial Credit Company in Baltimore. I found the same problems arising in business that had arisen during the war: a score of things had to be done at once-and there was little time to do them. We were low in stocks. We had new forms to handle, new stock arrangements, changes of address, opening and closing offices, and so on. Instead of getting taut and nervous, I remembered what the doctor had told me. 'One grain of sand at a time. One task at a time.' By repeating those words to myself over and over, I accomplished my tasks in a more efficient manner and I did my work without the confused and jumbled feeling that had almost wrecked me on the battlefield."

One of the most appalling comments on our present way of life is that half of all the beds in our hospitals are reserved for patients with nervous and mental troubles, patients who have collapsed under the crushing burden of accumulated yesterdays and fearful tomorrows. Yet a vast majority of those people would be walking the streets today, leading happy, useful lives, if they had only heeded the words of Jesus: "Have no anxiety about the morrow"; or the words of Sir William Osier: "Live in day-tight compartments".

You and I are standing this very second at the meeting-place of two eternities: the vast past that has endured forever, and the future that is plunging on to the last syllable of recorded time. We can't possibly live in either of those eternities-no, not even for one split second. But, by trying to do so, we can wreck both our bodies and our minds. So let's be content to live the only time we can possibly live: from now until bedtime. "Anyone can carry his burden, however hard, until nightfall," wrote Robert Louis Stevenson. "Anyone can do his work, however hard, for one day. Anyone can live sweetly, patiently, lovingly, purely, till the sun goes down. And this is all that life really means."

Yes, that is all that life requires of us; but Mrs. E. K. Shields, 815, Court Street, Saginaw, Michigan, was driven to despair- even to the brink of suicide-before she learned to live just till bedtime. "In 1937, I lost my husband," Mrs. Shields said as she told me her story. "I was very depressed-and almost penniless. I wrote my former employer, Mr. Leon Roach, of the Roach-Fowler Company of Kansas City, and got my old job back. I had formerly made my living selling books to rural and town school boards. I had sold my car two years previously when my husband became ill; but I managed to scrape together enough money to put a down payment on a used car and started out to sell books again.

"I had thought that getting back on the road would help relieve my depression; but driving alone and eating alone was almost more than I could take. Some of the territory was not very productive, and I found it hard to make those car payments, small as they were.

"In the spring of 1938, I was working out from Versailles, Missouri. The schools were poor, the roads bad; I was so lonely and discouraged that at one time I even considered suicide. It seemed that success was impossible. I had nothing to live for. I dreaded getting up each morning and facing life. I was afraid of everything: afraid I could not meet the car payments; afraid I could not pay my room rent; afraid I would not have enough to eat. I was afraid my health was failing and I had no money for a doctor. All that kept me from suicide were the thoughts that my sister would be deeply grieved, and that I did not have enough money to pay my funeral expenses.

"Then one day I read an article that lifted me out of my despondence and gave me the courage to go on living. I shall never cease to be grateful for one inspiring sentence in that article. It said: 'Every day is a new life to a wise man.' I typed that sentence out and pasted it on the windshield of my car, where I saw it every minute I was driving. I found it wasn't so hard to live only one day at a time. I learned to forget the yesterdays and to not-think of the tomorrows. Each morning I said to myself: 'Today is a new life.'

"I have succeeded in overcoming my fear of loneliness, my fear of want. I am happy and fairly successful now and have a lot of enthusiasm and love for life. I know now that I shall never again be afraid, regardless of what life hands me. I know now that I don't have to fear the future. I know now that I can live one day at a time-and that 'Every day is a new life to a wise man.'"

Who do you suppose wrote this verse:

Happy the man, and happy he alone,
He, who can call to-day his own:
He who, secure within, can say:
"To-morrow, do thy worst, for I have liv'd to-day."

Those words sound modern, don't they? Yet they were written thirty years before Christ was born, by the Roman poet Horace.

One of the most tragic things I know about human nature is that all of us tend to put off living. We are all dreaming of some magical rose garden over the horizon-instead of enjoying the roses that are blooming outside our windows today.

Why are we such fools-such tragic fools?

"How strange it is, our little procession of life! " wrote Stephen Leacock. "The child says: 'When I am a big boy.' But what is that? The big boy says: 'When I grow up.' And then, grown up, he says: 'When I get married.' But to be married, what is that after all? The thought changes to 'When I'm able to retire." And then, when retirement comes, he looks back over the landscape traversed; a cold wind seems to sweep over it; somehow he has missed it all, and it is gone. Life, we learn too late, is in the living, in the tissue of every day and hour."

The late Edward S. Evans of Detroit almost killed himself with worry before he learned that life "is in the living, in the tissue of every day and hour." Brought up in poverty, Edward Evans made his first money by selling newspapers, then worked as a grocer's clerk. Later, with seven people dependent upon him for bread and butter, he got a job as an assistant librarian. Small as the pay was, he was afraid to quit. Eight years passed before he could summon up the courage to start out on his own. But once he started, he built up an original investment of fifty-five borrowed dollars into a business of his own that made him twenty thousand dollars a year. Then came a frost, a killing frost. He endorsed a big note for a friend-and the friend went bankrupt.

Quickly on top of that disaster came another: the bank in which he had all his money collapsed. He not only lost every cent he had, but was plunged into debt for sixteen thousand dollars.

His nerves couldn't take it. "I couldn't sleep or eat," he told me. "I became strangely ill. Worry and nothing but worry," he said, "brought on this illness. One day as I was walking down the street, I fainted and fell on the sidewalk. I was no longer able to walk. I was put to bed and my body broke out in boils. These boils turned inward until just lying in bed was agony. I grew weaker every day. Finally my doctor told me that I had only two more weeks to live. I was shocked. I drew up my will, and then lay back in bed to await my end. No use now to struggle or worry. I gave up, relaxed, and went to sleep. I hadn't slept two hours in succession for weeks; but now with my earthly problems drawing to an end, I slept like a baby. My exhausting weariness began to disappear. My appetite returned. I gained weight.

"A few weeks later, I was able to walk with crutches. Six weeks later, I was able to go back to work. I had been making twenty thousand dollars a year; but I was glad now to get a job for thirty dollars a week. I got a job selling blocks to put behind the wheels of automobiles when they are shipped by freight. I had learned my lesson now. No more worry for me-no more regret about what had happened in the past- no more dread of the future. I concentrated all my time, energy, and enthusiasm into selling those blocks."

Edward S. Evans shot up fast now. In a few years, he was president of the company. His company-the Evans Product Company-has been listed on the New York Stock Exchange for years. When Edward S. Evans died in 1945, he was one of the most progressive businessmen in the United States. If you ever fly over Greenland, you may land on Evans Field- a flying-field named in his honour.

Here is the point of the story: Edward S. Evans would never have had the thrill of achieving these victories in business and in living if he hadn't seen the folly of worrying-if he hadn't learned to live in day-tight compartments.

Five hundred years before Christ was born, the Greek philosopher Heraclitus told his students that "everything changes except the law of change". He said: "You cannot step in the same river twice." The river changes every second; and so does the man who stepped in it. Life is a ceaseless change. The only certainty is today. Why mar the beauty of living today by trying to solve the problems of a future that is shrouded in ceaseless change and uncertainty in a future that no one can possibly foretell?

The old Romans had a word for it. In fact, they had two words for it. "*Carpe diem*". "Enjoy the day." Or, "Seize the day." Yes, seize the day, and make the most of it.

That is the philosophy of Lowell Thomas. I recently spent a weekend at his farm; and I noticed that he had these words from Psalm CXVIII framed and hanging on the walls of his broadcasting studio where he would see them often:

This is the day which the Lord hath made; we will rejoice and be glad in it.

John Ruskin had on his desk a simple piece of stone on which was carved one word: TODAY. And while I haven't a piece of stone on my desk, I do have a poem pasted on my mirror where I can see it when I shave every morning-a poem that Sir William Osier always kept on his desk-a poem written by the famous Indian dramatist, Kalidasa:

SALUTATION TO THE DAWN

Look to this day!
For it is life, the very life of life.

In its brief course
Lie all the verities and realities of your existence:
The bliss of growth
The glory of action
The splendour of achievement.

For yesterday is but a dream
And tomorrow is only a vision,
But today well lived makes yesterday a dream of happiness
And every tomorrow a vision of hope.

Look well, therefore, to this day!
Such is the salutation to the dawn.

So, the first thing you should know about worry is this: if you want to keep it out of your life, do what Sir William Osier did -

1.1

RULE 1

Shut the iron doors on the past and the future. Live in Daytight Compartments

Why not ask yourself these questions, and write down the answers?

1. Do I tend to put off living in the present in order to worry about the future, or to yearn for some "magical rose garden over the horizon"?
2. Do I sometimes embitter the present by regretting things that happened in the past-that are over and done with?
3. Do I get up in the morning determined to "Seize the day"-to get the utmost out of these twenty-four hours?
4. Can I get more out of life by "living in day-tight compartments"?
5. When shall I start to do this? Next week? .. Tomorrow? ... Today?

* * *

Chapter 2

A Magic Formula For Solving Worry Situations

Would you like a quick, sure-fire recipe for handling worry situations-a technique you can start using right away, before you go any further in reading this book?

Then let me tell you about the method worked out by Willis H. Carrier, the brilliant engineer who launched the air-conditioning industry, and who is now head of the world-famous Carrier Corporation in Syracuse, New York. It is one of the best techniques I ever heard of for solving worry problems, and I got it from Mr. Carrier personally when we were having lunch together one day at the Engineers' Club in New York.

"When I was a young man," Mr. Carrier said, "I worked for the Buffalo Forge Company in Buffalo, New York. I was handed the assignment of installing a gas-cleaning device in a plant of the Pittsburgh Plate Glass Company at Crystal City, Missouri-a plant costing millions of dollars. The purpose of this installation was to remove the impurities from the gas so it could be burned without injuring the engines. This method of cleaning gas was new. It had been tried only once before- and under different conditions. In my work at Crystal City, Missouri, unforeseen difficulties arose. It worked after a fashion -but not well enough to meet the guarantee we had made.

"I was stunned by my failure. It was almost as if someone had struck me a blow on the head. My stomach, my insides, began to twist and turn. For a while I was so worried I couldn't sleep.

"Finally, common sense reminded me that worry wasn't getting me anywhere; so I figured out a way to handle my problem without worrying. It worked superbly. I have been using this same anti-worry technique for more than thirty years.

"It is simple. Anyone can use it. It consists of three steps:

"Step I. I analysed the situation fearlessly and honestly and figured out what was the worst that could possibly happen as a result of this failure. No one was going to jail me or shoot me. That was certain. True, there was a chance that I would lose my position; and there was also a chance that my employers would have to remove the machinery and lose the twenty thousand dollars we had invested.

"Step II. After figuring out what was the worst that could possibly happen, I reconciled myself to accepting it, if necessary. I said to myself: This failure will be a blow to my record, and it might possibly mean the loss of my job; but if it does, I can always get another position. Conditions could be much worse; and as far as my employers are concerned- well, they realise that we are experimenting with a new method of cleaning gas, and if this experience costs them twenty thousand dollars, they can stand it. They can charge it up to research, for it is an experiment.

"After discovering the worst that could possibly happen and reconciling myself to accepting it, if necessary, an extremely important thing happened: I immediately relaxed and felt a sense of peace that I hadn't experienced in days.

"Step III. From that time on, I calmly devoted my time and energy to trying to improve upon the worst which I had already accepted mentally.

"I now tried to figure out ways and means by which I might reduce the loss of twenty thousand dollars that we faced. I made several tests and finally figured out that if we spent another five thousand for additional equipment, our problem would be solved. We did this, and instead of the firm losing twenty thousand, we made fifteen thousand.

"I probably would never have been able to do this if I had kept on worrying, because one of the worst features about worrying is that it destroys our ability to concentrate. When we worry, our minds jump here and there and everywhere, and we lose all power of decision. However, when we force ourselves to face the worst and accept it mentally, we then eliminate all those vague imaginings and put ourselves in a position in which we are able to concentrate on our problem.

"This incident that I have related occurred many years ago. It worked so superbly that I have been using it ever since; and, as a result, my life has been almost completely free from worry."

Now, why is Willis H. Carrier's magic formula so valuable and so practical, psychologically speaking? Because it yanks us down out of the great grey clouds in which we fumble around when we are blinded by worry. It plants our feet good and solid on the earth. We know where we stand. And if we haven't solid ground under us, how in creation can we ever hope to think anything through?

Professor William James, the father of applied psychology, has been dead for thirty-eight years. But if he were alive today, and could hear his formula for facing the worst, he would heartily approve it. How do I know that? Because he told his own students: "Be willing to have it so .... Be willing to have it so," he said, because "... Acceptance of what has happened is the first step in overcoming the consequences of any misfortune."

The same idea was expressed by Lin Yutang in his widely read book, *The Importance of Living*. "True peace of mind," said this Chinese philosopher, "comes from accepting the worst. Psychologically, I think, it means a release of energy."

That's it, exactly! Psychologically, it means a new release of energy! When we have accepted the worst, we have nothing more to lose. And that automatically means-we have everything to gain! "After facing the worst," Willis H. Carrier reported, "I immediately relaxed and felt a sense of peace that I hadn't experienced in days. From that time on, I was able to think."

Makes sense, doesn't it? Yet millions of people have wrecked their lives in angry turmoil, because they refused to accept the worst; refused to try to improve upon it; refused to salvage what they could from the wreck. Instead of trying to reconstruct their fortunes, they engaged in a bitter and "violent contest with experience"-and ended up victims of that brooding fixation known as melancholia.

Would you like to see how someone else adopted Willis H. Carrier's magic formula and applied it to his own problem? Well, here is one example, from a New York oil dealer who was a student in my classes.

"I was being blackmailed!" this student began. "I didn't believe it was possible-I didn't believe it could happen outside of the movies-but I was actually being blackmailed! What happened was this: the oil company of which I was the head had a number of delivery trucks and a number of drivers. At that time, OPA regulations were strictly in force, and we were rationed on the amount of oil we could deliver to any one of our customers. I didn't know it, but it seems that certain of our drivers had been delivering oil short to our regular customers, and then reselling the surplus to customers of their own.

"The first inkling I had of these illegitimate transactions was when a man who claimed to be a government inspector came to see me one day and demanded hush money. He had got documentary proof of what our drivers had been doing, and he threatened to turn this proof over to the District Attorney's office if I didn't cough up.

"I knew, of course, that I had nothing to worry about-personally, at least. But I also knew that the law says a firm is responsible for the actions of its employees. What's more, I knew that if the case came to court, and it was aired in the newspapers, the bad publicity would ruin my business. And I was proud of my business-it had been founded by my father twenty-four years before.

"I was so worried I was sick! I didn't eat or sleep for three days and nights. I kept going around in crazy circles. Should I pay the money-five thousand dollars-or should I tell this man to go ahead and do his damnedest? Either way I tried to make up my mind, it ended in nightmare.

"Then, on Sunday night, I happened to pick up the booklet on How to Stop Worrying which I had been given in my

Carnegie class in public speaking. I started to read it, and came across the story of Willis H. Carrier. 'Face the worst', it said. So I asked myself: 'What is the worst that can happen if I refuse to pay up, and these blackmailers turn their records over to the District Attorney?'

"The answer to that was: 'The ruin of my business-that's the worst that can happen. I can't go to jail. All that can happen is that I shall be ruined by the publicity.'

"I then said to myself: 'All right, the business is ruined. I accept that mentally. What happens next?'

"Well, with my business ruined, I would probably have to look for a job. That wasn't bad. I knew a lot about oil- there were several firms that might be glad to employ me. ... I began to feel better. The blue funk I had been in for three days and nights began to lift a little. My emotions calmed down. ... And to my astonishment, I was able to think.

"I was clear-headed enough now to face Step III-improve on the worst. As I thought of solutions, an entirely new angle presented itself to me. If I told my attorney the whole situation, he might find a way out which I hadn't thought of. I know it sounds stupid to say that this hadn't even occurred to me before-but of course I hadn't been thinking, I had only been worrying! I immediately made up my mind that I would see my attorney first thing in the morning-and then I went to bed and slept like a log!

"How did it end? Well, the next morning my lawyer told me to go and see the District Attorney and tell him the truth. I did precisely that. When I finished I was astonished to hear the D.A. say that this blackmail racket had been going on for months and that the man who claimed to be a 'government agent' was a crook wanted by the police. What a relief to hear all this after I had tormented myself for three days and nights wondering whether I should hand over five thousand dollars to this professional swindler!

"This experience taught me a lasting lesson. Now, whenever I face a pressing problem that threatens to worry me, I give it what I call 'the old Willis H. Carrier formula'."

At just about the same time Willis H. Carrier was worrying over the gas-cleaning equipment he was installing in a plant in Crystal City, Missouri, a chap from Broken Bow, Nebraska, was making out his will. His name was Earl P. Haney, and he had duodenal ulcers. Three doctors, including a celebrated ulcer specialist, had pronounced Mr. Haney an "incurable case". They had told him not to eat this or that, and not to worry or fret-to keep perfectly calm. They also told him to make out his will!

These ulcers had already forced Earl P. Haney to give up a fine and highly paid position. So now he had nothing to do, nothing to look forward to except a lingering death.

Then he made a decision: a rare and superb decision. "Since I have only a little while to live," he said, "I may as well make the most of it. I have always wanted to travel around the world before I die. If I am ever going to do it, I'll have to do it now." So he bought his ticket.

The doctors were appalled. "We must warn you," they said to Mr. Haney, "that if you do take this trip, you will be buried at sea."

"No, I won't," he replied. "I have promised my relatives that I will be buried in the family plot at Broken Bow, Nebraska. So I am going to buy a casket and take it with me."

He purchased a casket, put it aboard ship, and then made arrangements with the steamship company in the event of his death-to put his corpse in a freezing compartment and keep it there till the liner returned home. He set out on his trip, imbued with the spirit of old Omar:

Ah, make the most of what we yet may spend,
Before we too into the Dust descend;
Dust into Dust, and under Dust, to lie,
Sans Wine, sans Song, sans Singer, and-sans End!

However, he didn't make the trip "sans wine". "I drank highballs, and smoked long cigars on that trip," Mr. Haney says in a letter that I have before me now. "I ate all kinds of foods-even strange native foods which were guaranteed to kill me. I enjoyed myself more than I had in years! We ran into monsoons and typhoons which should have put me in my casket, if only from fright-but I got an enormous kick out of all this adventure.

"I played games aboard the ship, sang songs, made new friends, stayed up half the night. When we reached China and India, I realised that the business troubles and cares that I had faced back home were paradise compared to the poverty and hunger in the Orient. I stopped all my senseless worrying and felt fine. When I got back to America, I had gained ninety pounds. I had almost forgotten I had ever had a stomach ulcer. I had never felt better in my life. I promptly sold the casket back to the undertaker, and went back to business. I haven't been ill a day since."

At the time this happened, Earl P. Haney had never even heard of Willis H. Carrier and his technique for handling worry. "But I realise now," he told me quite recently, "that I was unconsciously using the selfsame principle. I reconciled myself to the worst that could happen-in my case, dying. And then I improved upon it by trying to get the utmost enjoyment out of life for the time I had left. ... If," he continued, "if I had gone on worrying after boarding that ship, I have no doubt that I would have made the return voyage inside of that coffin. But I relaxed-I forgot it. And this calmness of mind gave me a new birth of energy which actually saved my life." (Earl P. Haney is now living at 52 Wedgemere Ave., Winchester, Mass.)

Now, if Willis H. Carrier could save a twenty-thousand-dollar contract, if a New York businessman could save himself from blackmail, if Earl P. Haney could actually save his life, by using this magic formula, then isn't it possible that it may be the answer to some of your troubles? Isn't it possible that it may even solve some problems you thought were unsolvable?

So, Rule 2 is

If you have a worry problem, apply the magic formula of Willis H. Carrier by doing these three things-

1.2

RULE 2

1. Ask yourself, 'What is the worst that can possibly happen?'
2. Prepare to accept it if you have to.
3. Then calmly proceed to improve on the worst.

✤ ✤ ✤

Chapter 3

What Worry May Do To You

Businessmen who do not know how to fight worry die young.

-Dr. Alexis Carrel.

Some time ago, a neighbour rang my doorbell one evening and urged me and my family to be vaccinated against smallpox. He was only one of thousands of volunteers who were ringing doorbells all over New York City. Frightened people stood in lines for hours at a time to be vaccinated. Vaccination stations were opened not only in all hospitals, but also in fire-houses, police precincts, and in large industrial plants. More than two thousand doctors and nurses worked feverishly day and night, vaccinating crowds. The cause of all this excitement? Eight people in New York City had smallpox-and two had died. Two deaths out of a population of almost eight million.

Now, I have lived in New York for over thirty-seven years, and no one has ever yet rung my doorbell to warn me against the emotional sickness of worry-an illness that, during the last thirty-seven years, has caused ten thousand times more damage than smallpox.

No doorbell ringer has ever warned me that one person out of ten now living in these United States will have a nervous breakdown-induced in the vast majority of cases by worry and emotional conflicts. So I am writing this chapter to ring your doorbell and warn you.

The great Nobel prizewinner in medicine, Dr. Alexis Carrel, said: "Businessmen who do not know how to fight worry die young." And so do housewives and horse doctors and bricklayers.

A few years ago, I spent my vacation motoring through Texas and New Mexico with Dr. O. F. Gober-one of the medical executives of the Santa Fe railway. His exact title was chief physician of the Gulf, Colorado and Santa Fe Hospital Association. We got to talking about the effects of worry, and he said: "Seventy per cent of all patients who come to physicians could cure themselves if they only got rid of their fears and worries. Don't think for a moment that I mean that their ills are imaginary," he said. "Their ills are as real as a throbbing toothache and sometimes a hundred times more serious. I refer to such illnesses as nervous indigestion, some stomach ulcers, heart disturbances, insomnia, some headaches, and some types of paralysis.

"These illnesses are real. I know what I am talking about," said Dr. Gober, "for I myself suffered from a stomach ulcer for twelve years.

"Fear causes worry. Worry makes you tense and nervous and affects the nerves of your stomach and actually changes the gastric juices of your stomach from normal to abnormal and often leads to stomach ulcers."

Dr. Joseph F. Montague, author of the book *Nervous Stomach Trouble*, says much the same thing. He says: "You do not get stomach ulcers from what you eat. You get ulcers from what is eating you."

Dr. W.C. Alvarez, of the Mayo Clinic, said: "Ulcers frequently flare up or subside according to the hills and valleys of emotional stress."

That statement was backed up by a study of 15,000 patients treated for stomach disorders at the Mayo Clinic. Four out of five had no physical basis whatever for their stomach illnesses. Fear, worry, hate, supreme selfishness, and the inability to adjust themselves to the world of reality-these were largely the causes of their stomach illnesses and stomach ulcers. ... Stomach ulcers can kill you. According to Life magazine, they now stand tenth in our list of fatal diseases.

I recently had some correspondence with Dr. Harold C. Habein of the Mayo Clinic. He read a paper at the annual meeting of the American Association of Industrial Physicians and Surgeons, saying that he had made a study of 176 business executives whose average age was 44.3 years. He reported that slightly more than a third of these executives suffered from one of three ailments peculiar to hightension living-heart disease, digestive-tract ulcers, and high blood pressure. Think of it- a third of our business executives are wrecking their bodies with heart disease, ulcers, and high blood pressure before they even reach forty-five. What price success! And they aren't even buying success! Can any man possibly be a success who is paying for business advancement with stomach ulcers and heart trouble? What shall it profit a man if he gains the whole world-and loses his health? Even if he owned the whole world, he could sleep in only one bed at a time and eat only three meals a day. Even a ditchdigger can do that-and probably sleep more soundly and enjoy his food more than a high-powered executive. Frankly, I would rather be a share-cropper down in Alabama with a banjo on my knee than wreck my health at forty-five by trying to run a railroad or a cigarette company. And speaking of cigarettes-the best-known cigarette manufacturer in the world recently dropped dead from heart failure while trying to take a little recreation in the Canadian woods. He amassed millions and fell dead at sixty-one. He probably traded years of his life for what is called "business success".

In my estimation, this cigarette executive with all his millions was not half as successful as my father - a Missouri farmer- who died at eighty-nine without a dollar.

The famous Mayo brothers declared that more than half of our hospital beds are occupied by people with nervous troubles. Yet, when the nerves of these people are studied under a high-powered microscope in a post-mortem examination, their nerves in most cases are apparently as healthy as the nerves of Jack Dempsey. Their "nervous troubles" are caused not by a physical deterioration of the nerves, but by emotions of futility, frustration, anxiety, worry, fear, defeat, despair. Plato said that "the greatest mistake physicians make is that they attempt to cure the body without attempting to cure the mind; yet the mind and body are one and should not be treated separately!"

It took medical science twenty-three hundred years to recognise this great truth. We are just now beginning to develop a new kind of medicine called psychosomatic medicine-a medicine that treats both the mind and the body. It is high time we were doing that, for medical science has largely wiped out the terrible diseases caused by physical germs--diseases such as smallpox, cholera, yellow fever, and scores of other scourges that swept untold millions into untimely graves. But medical science has been unable to cope with the mental and physical wrecks caused, not by germs, but by emotions of worry, fear, hate, frustration, and despair. Casualties caused by these emotional diseases are mounting and spreading with catastrophic rapidity.

Doctors figure that one American in every twenty now alive will spend a part of his life in an institution for the mentally ill. One out of every six of our young men called up by the draft in the Second World War was rejected as mentally diseased or defective.

What causes insanity? No one knows all the answers. But it is highly probable that in many cases fear and worry are contributing factors. The anxious and harassed individual who is unable to cope with the harsh world of reality breaks off all contact with his environment and retreats into a private dream world of his own making, and this solves his worry problems.

As I write I have on my desk a book by Dr. Edward Podolsky entitled *Stop Worrying and Get Well.*

Here are some of the chapter titles in that book:
What Worry Does To The Heart
High Blood Pressure Is Fed By Worry
Rheumatism Can Be Caused By Worry
Worry Less For Your Stomach's Sake
How Worry Can Cause A Cold
Worry And The Thyroid
The Worrying Diabetic

Another illuminating book about worry is *Man Against Himself*, by Dr. Karl Menninger, one of the "Mayo brothers of psychiatry." Dr. Menninger's book is a startling revelation of what you do to yourself when you permit destructive emotions to dominate your life. If you want to stop working against yourself, get this book. Read it. Give it to your friends. It costs four dollars-and is one of the best investments you can make in this life.

Worry can make even the most stolid person ill. General Grant discovered that during the closing days of the Civil War. The story goes like this: Grant had been besieging Richmond for nine months. General Lee's troops, ragged and hungry, were beaten. Entire regiments were deserting at a time. Others were holding prayer meetings in their tents-shouting, weeping, seeing visions. The end was close. Lee's men set fire to the cotton and tobacco warehouses in Richmond, burned the arsenal, and fled from the city at night while towering flames roared up into darkness. Grant was in hot pursuit, banging away at the Confederates from both sides and the rear, while Sheridan's cavalry was heading them off in front, tearing up railway lines and capturing supply trains.

Grant, half blind with a violent sick headache, fell behind his army and stopped at a farmhouse. "I spent the night," he records in his Memoirs, "in bathing my feet in hot water and mustard, and putting mustard plasters on my wrists and the back part of my neck, hoping to be cured by morning."

The next morning, he was cured instantaneously. And the tiling that cured him was not a mustard plaster, but a horseman galloping down the road with a letter from Lee, saying he wanted to surrender.

"When the officer [bearing the message] reached me," Grant wrote, "I was still suffering with the sick headache, but the instant I saw the contents of the note, I was cured."

Obviously it was Grant's worries, tensions, and emotions that made him ill. He was cured instantly the moment his emotions took on the hue of confidence, achievement, and victory.

Seventy years later, Henry Morgenthau, Jr., Secretary of the Treasury in Franklin D. Roosevelt's cabinet, discovered that worry could make him so ill that he was dizzy. He records in his diary that he was terribly worried when the President, in order to raise the price of wheat, bought 4,400,000 bushels in one day. He says in his diary: "I felt literally dizzy while the thing was going on. I went home and went to bed for two hours after lunch."

If I want to see what worry does to people, I don't have to go to a library or a physician. I can look out of the window of my home where I am writing this book; and I can see, within one block, one house where worry caused a nervous breakdown-and another house where a man worried himself into diabetes. When the stock market went down, the sugar in his blood and urine went up.

When Montaigne, the illustrious French philosopher, was elected Mayor of his home town-Bordeaux he said to his fellow citizens: "I am willing to take your affairs into my hands but not into my liver and lungs."

This neighbour of mine took the affairs of the stock market into the blood stream-and almost killed himself.

Worry can put you into a wheel chair with rheumatism and arthritis. Dr. Russell L. Cecil, of the Cornell University Medical School, is a world-recognised authority on arthritis; and he has listed four of the commonest conditions that bring on arthritis:

1. Marital shipwreck.
2. Financial disaster and grief.
3. Loneliness and worry.
4. Long-cherished resentments.

Naturally, these four emotional situations are far from being the only causes of arthritis. There are many different kinds of arthritis-due to various causes. But, to repeat, the commonest conditions that bring on arthritis are the four listed by Dr. Russell L. Cecil. For example, a friend of mine was so hard hit during the depression that the gas company shut off the gas and the bank foreclosed the mortgage on the house. His wife suddenly had a painful attack of arthritis-and, in spite of medicine and diets, the arthritis continued until their financial situation improved.

Worry can even cause tooth decay. Dr. William I.L. McGonigle said in an address before the American Dental Association that "unpleasant emotions such as those caused by worry, fear, nagging ... may upset the body's calcium balance and cause tooth decay". Dr. McGonigle told of a patient of his who had always had a perfect set of teeth until he began to worry over his wife's sudden illness. During the three weeks she was in the hospital, he developed nine cavities- cavities brought on by worry.

Have you ever seen a person with an acutely over-active thyroid? I have, and I can tell you they tremble; they shake; they look like someone half scared to death-and that's about what it amounts to. The thyroid gland, the gland that regulates the body, has been thrown out of kilter. It speeds up the heart -the whole body is roaring away at full blast like a furnace with all its draughts wide open. And if this isn't checked, by operation or treatment, the victim may die, may "burn himself out".

A short time ago I went to Philadelphia with a friend of mine who has this disease. We went to see a famous specialist, a doctor who has been treating this type of ailment for thirty-eight years. And what sort of advice do you suppose he had hanging on the wall of his waiting-room-painted on a large wooden sign so all his patients could see it? Here it is. I copied it down on the back of an envelope while I was waiting:

Relaxation and Recreation

> The most relaxing recreating forces are a healthy
> religion, sleep, music, and laughter.
> Have faith in God-learn to sleep well-
> Love good music-see the funny side of life-
> And health and happiness will be yours.

The first question he asked this friend of mine was: "What emotional disturbance brought on this condition?" He warned my friend that, if he didn't stop worrying, he could get other complications: heart trouble, stomach ulcers, or diabetes. "All of these diseases," said that eminent doctor, "are cousins, first cousins." Sure, they're first cousins-they're all worry diseases!

When I interviewed Merle Oberon, she told me that she refused to worry because she knew that worry would destroy her chief asset on the motion-picture screen: her good looks.

"When I first tried to break into the movies," she told me, "I was worried and scared. I had just come from India, and I didn't know anyone in London, where I was trying to get a job. I saw a few producers, but none of them hired me; and the little money I had began to give out. For two weeks I lived on nothing but crackers and water. I was not only worried now. I was hungry. I said to myself: 'Maybe you're a fool. Maybe you will never break into the movies. After all, you have no experience, you've never acted at all-what have you to offer but a rather pretty face?'

"I went to the mirror. And when I looked in that mirror, I saw what worry was doing to my looks! I saw the lines it was forming. I saw the anxious expression. So I said to myself: 'You've got to stop this at once! You can't afford to worry. The only thing you have to offer at all is your looks, and worry will ruin them I'"

Few things can age and sour a woman and destroy her looks as quickly as worry. Worry curdles the expression. It makes us clench our jaws and lines our faces with wrinkles. It forms a permanent scowl. It may turn the hair grey, and in some cases, even make it fall out. It can ruin the complexion- it can bring on all kinds of skin rashes, eruptions, and pimples.

Heart disease is the number-one killer in America today. During the Second World War, almost a third of a million men were killed in combat; but during that same period, heart disease killed two million civilians-and one million of those casualties were caused by the kind of heart disease that is brought on by worry and high-tension living. Yes, heart disease is one of the chief reasons why Dr. Alexis Carrel said: "Businessmen who do not know how to fight worry die young."

The Negroes down south and the Chinese rarely have the kind of heart disease brought on by worry, because they take things calmly. Twenty times as many doctors as farm workers die from heart failure. The doctors lead tense lives-and pay the penalty.

"The Lord may forgive us our sins," said William James, "but the nervous system never does." Here is a startling and almost incredible fact: more Americans commit suicide each year than die from the five most common communicable diseases.

Why? The answer is largely: "Worry."

When the cruel Chinese war lords wanted to torture their prisoners, they would tie their prisoners hand and foot and put them under a bag of water that constantly dripped ... dripped ... dripped ... day and night. These drops of water constantly falling on the head finally became like the sound of hammer blows-and drove men insane. This same method of torture was used during the Spanish Inquisition and in German concentration camps under Hitler.

Worry is like the constant drip, drip, drip of water; and the constant drip, drip, drip of worry often drives men to insanity and suicide.

When I was a country lad in Missouri, I was half scared to death by listening to Billy Sunday describe the hell-fires of the next world. But he never ever mentioned the hell-fires of physical agony that worriers may have here and now. For example, if you are a chronic worrier, you may be stricken some day with one of the most excruciating pains ever endured by man: angina pectoris.

Boy, if that ever hits you, you will scream with agony. Your screams will make the sounds in Dante's Inferno sound like Babes in Toyland. You will say to yourself then: "Oh, God, oh, God, if I can ever get over this, I will never worry about anything-ever." (If you think I am exaggerating, ask your family physician.)

Do you love life? Do you want to live long and enjoy good health? Here is how you can do it. I am quoting Dr. Alexis Carrel again. He said: "Those who keep the peace of their inner selves in the midst of the tumult of the modern city are immune from nervous diseases."

Can you keep the peace of your inner self in the midst of the tumult of a modem city? If you are a normal person, the answer is "yes". "Emphatically yes." Most of us are stronger than we realise. We have inner resources that we have probably never tapped. As Thoreau said in his immortal book, Walden:

"I know of no more encouraging fact than the unquestionable ability of man to elevate his life by a conscious endeavour. ... If one advances confidently in the direction of his dreams, and endeavours to live the life he has imagined, he will meet with a success unexpected in common hours."

Surely, many of the readers of this book have as much will power and as many inner resources as Olga K. Jarvey has. Her address is Box 892, Coeur d'Alene, Idaho. She discovered that under the most tragic circumstances she could banish worry. I firmly believe that you and I can also-if we apply the old, old truths discussed in this volume. Here is Olga K. Jarvey's story as she wrote it for me: "Eight and a half years ago, I was condemned to die-a slow, agonising death-of cancer. The best medical brains of the country, the Mayo brothers, confirmed the sentence. I was at a dead-end street, the ultimate gaped at me! I was young. I did not want to die! In my desperation, I phoned to my doctor at Kellogg and cried out to him the despair in my heart. Rather impatiently he upbraided me: 'What's the matter, Olga, haven't you any fight in you? Sure, you will die if you keep on crying. Yes, the worst has overtaken you. O.K.-face the facts! Quit worrying! And then do something about it!' Right then and there I took an oath, an oath so solemn that the nails sank deep into my flesh and cold chills ran down my spine: 'I am not going to worry! I am not going to cry! And if there is anything to mind over matter, I am going to win! I am going to LIVE!'

"The usual amount of X-ray in such advanced cases, where they cannot apply radium, is 10 1/2 minutes a day for 30 days. They gave me X-ray for 14 1/2 minutes a day for 49 days; and although my bones stuck out of my emaciated body like rocks on a barren hillside, and although my feet were like lead, I did not worry! Not once did I cry! I smiled! Yes, I actually forced myself to smile.

"I am not so foolish as to imagine that merely smiling can cure cancer. But I do believe that a cheerful mental attitude helps the body fight disease. At any rate, I experienced one of the miracle cures of cancer. I have never been healthier than in the last few years, thanks to those challenging, fighting words of Dr. McCaffery: 'Face the facts: Quite worrying; then do something about it!'"

I am going to close this chapter by repeating its title: the words of Dr. Alexis Carrel: "Businessmen who do not know how to fight worry die young."

The fanatical followers of the prophet Mohammed often had verses from the Koran tattooed on their breasts. I would like to have the title of this chapter tattooed on the breast of every reader of this book: "Businessmen who do not know how to fight worry die young."

Was Dr. Carrel speaking of you?

Could be.

Part One In A Nutshell

RULE 1 If you want to avoid worry, do what Sir William Osier did: Live in "day-tight compartments". Don't stew about the future. Just live each day until bedtime.

RULE 2 The next time Trouble-with a capital T- comes gunning for you and backs you up in a corner, try the magic formula of Willis H. Carrier:

a. Ask yourself, "What is the worst that can possibly happen if I can't solve my problem?"

b. Prepare yourself mentally to accept the worst-if necessary.
c. Then calmly try to improve upon the worst-which you have already mentally agreed to accept.

RULE 3 Remind yourself of the exorbitant price you can pay for worry in terms of your health. "Businessmen who do not know how to fight worry die young."

✦ ✦ ✦

PART TWO

Basic Techniques In Analysing Worry

Chapter 4

How To Analyse And Solve Worry Problems

I keep six honest serving-men
(They taught me all I knew):
Their names are What and Why and When
And How and Where and Who.

-Rudyard Kipling

Will the magic formula of Willis H. Carrier, described in Part One, Chapter 2, solve all worry problems? No, of course not. Then what is the answer? The answer is that we must equip ourselves to deal with different kinds of worries by learning the three basic steps of problem analysis. The three steps are:

1. Get the facts.
2. Analyse the facts.
3. Arrive at a decision-and then act on that decision.

Obvious stuff? Yes, Aristotle taught it-and used it. And you and I must use it too if we are going to solve the problems that are harassing us and turning our days and nights into veritable hells.

Let's take the first rule: Get the facts. Why is it so important to get the facts? Because unless we have the facts we can't possibly even attempt to solve our problem intelligently. Without the facts, all we can do is stew around in confusion. My idea? No, that was the idea of the late Herbert E. Hawkes, Dean of Columbia College, Columbia University, for twenty-two years. He had helped two hundred thousand students solve their worry problems; and he told me that "confusion is the chief cause of worry". He put it this way-he said: "Half the worry in the world is caused by people trying to make decisions before they have sufficient knowledge on which to base a decision. For example," he said, "if I have a problem which has to be faced at three o'clock next Tuesday, I refuse even to try to make a decision about it until next Tuesday arrives. In the meantime, I concentrate on getting all the facts that bear on the problem. I don't worry," he said, "I don't agonise over my problem. I don't lose any sleep. I simply concentrate on getting the facts. And by the time Tuesday rolls around, if I've got all the facts, the problem usually solves itself!"

I asked Dean Hawkes if this meant he had licked worry entirely. "Yes," he said, "I think I can honestly say that my life is now almost totally devoid of worry. I have found," he went on, "that if a man will devote his time to securing facts in an impartial, objective way, his worries usually evaporate in the light of knowledge." Let me repeat that: "If a man will devote his time to securing facts in an impartial, objective way, his worries will usually evaporate in the light of knowledge."

But what do most of us do ? If we bother with facts at all-and Thomas Edison said in all seriousness: "There is no expedient to which a man will not resort to avoid the labour of thinking"-if we bother with facts at all, we hunt like bird dogs after the facts that bolster up what we already think-and ignore all the others! We want only the facts that justify our acts-the facts that fit in conveniently with our wishful thinking and justify our preconceived prejudices!

As Andre Maurois put it: "Everything that is in agreement with our personal desires seems true. Everything that is not puts us into a rage."

Is it any wonder, then, that we find it so hard to get at the answers to our problems? Wouldn't we have the same trouble trying to solve a second-grade arithmetic problem, if we went ahead on the assumption that two plus two equals five? Yet there are a lot of people in this world who make life a hell for themselves and others by insisting that two plus two equals five-or maybe five hundred!

What can we do about it? We have to keep our emotions out of our thinking; and, as Dean Hawkes put it, we must secure the facts in "an impartial, objective" manner.

That is not an easy task when we are worried. When we are worried, our emotions are riding high. But here are two ideas that I have found helpful when trying to step aside from my problems, in order to see the facts in a clear, objective manner.

1. When trying to get the facts, I pretend that I am collecting this information not for myself, but for some other person. This helps me to take a cold, impartial view of the evidence. This helps me eliminate my emotions.

2. While trying to collect the facts about the problem that is worrying me, I sometimes pretend that I am a lawyer preparing to argue the other side of the issue. In other words, I try to get all the facts against myself-all the facts that are damaging to my wishes, all the facts I don't like to face.

Then I write down both my side of the case and the other side of the case-and I generally find that the truth lies somewhere in between these two extremities.

Here is the point I am trying to make. Neither you nor I nor Einstein nor the Supreme Court of the United States is brilliant enough to reach an intelligent decision on any problem without first getting the facts. Thomas Edison knew that. At the time of his death, he had two thousand five hundred notebooks filled with facts about the problems he was facing.

So Rule 1 for solving our problems is: Get the facts. Let's do what Dean Hawkes did: let's not even attempt to solve our problems without first collecting all the facts in an impartial manner.

However, getting all the facts in the world won't do us any good until we analyse them and interpret them.

I have found from costly experience that it is much easier to analyse the facts after writing them down. In fact, merely writing the facts on a piece of paper and stating our problem clearly goes a long way toward helping us to reach a sensible decision. As Charles Kettering puts it: "A problem well stated is a problem half solved."

Let me show you all this as it works out in practice. Since the Chinese say one picture is worth ten thousand words, suppose I show you a picture of how one man put exactly what we are talking about into concrete action.

Let's take the case of Galen Litchfield-a man I have known for several years; one of the most successful American businessmen in the Far East. Mr. Litchfield was in China in 1942, when the Japanese invaded Shanghai. And here is his story as he told it to me while a guest in my home:

"Shortly after the Japs took Pearl Harbour," Galen Litchfield began, "they came swarming into Shanghai. I was the manager of the Asia Life Insurance Company in Shanghai. They sent us an 'army liquidator'-he was really an admiral- and gave me orders to assist this man in liquidating our assets. I didn't have any choice in the matter. I could co-operate-or else. And the 'or else' was certain death.

"I went through the motions of doing what I was told, because I had no alternative. But there was one block of securities, worth $750,000, which I left off the list I gave to the admiral. I left that block of securities off the list because they belonged to our Hong Kong organisation and had nothing to do with the Shanghai assets. All the same, I feared I might be in hot water if the Japs found out what I had done. And they soon found out.

"I wasn't in the office when the discovery was made, but my head accountant was there. He told me that the Jap admiral flew into a rage, and stamped and swore, and called me a thief and a traitor! I had defied the Japanese Army! I knew what that meant. I would be thrown into the Bridge House!

"The Bridge House! The torture chamber of the Japanese Gestapo! I had had personal friends who had killed themselves rather than be taken to that prison. I had had other friends who had died in that place after ten days of questioning and torture. Now I was slated for the Bridge House myself!

"What did I do? I heard the news on Sunday afternoon. I suppose I should have been terrified. And I would have been terrified if I hadn't had a definite technique for solving my problems. For years, whenever I was worried I had always gone to my typewriter and written down two questions-and the answers to these questions:

1. What am I worrying about?
2. What can I do about it?

"I used to try to answer those questions without writing them down. But I stopped that years ago. I found that writing down both the questions and the answers clarifies my thinking.

So, that Sunday afternoon, I went directly to my room at the Shanghai Y.M.C.A. and got out my typewriter. I wrote:

1. What am I worrying about?

I am afraid I will be thrown into the Bridge House tomorrow morning.

Then I typed out the second question:

2. What can I do about it?

I spent hours thinking out and writing down the four courses of action I could take-and what the probable consequence of each action would be.

1. I can try to explain to the Japanese admiral. But he "no speak English". If I try to explain to him through an interpreter, I may stir him up again. That might mean death, for he is cruel, would rather dump me in the Bridge House than bother talking about it.

2. I can try to escape. Impossible. They keep track of me all the time. I have to check in and out of my room at the Y.M.C.A. If I try to escape, I'll probably be captured and shot.

3. I can stay here in my room and not go near the office again. If I do, the Japanese admiral will be suspicion, will probably send soldiers to get me and throw me into the Bridge House without giving me a chance to say a word.

4. I can go down to the office as usual on Monday morning. If I do, there is a chance that the Japanese admiral may be so busy that he will not think of what I did. Even if he does think of it, he may have cooled off and may not bother me. If this

happens, I am all right. Even if he does bother me, I'll still have a chance to try to explain to him. So, going down to the office as usual on Monday morning, and acting as if nothing had gone wrong gives me two chances to escape the Bridge House.

As soon as I thought it all out and decided to accept the fourth plan-to go down to the office as usual on Monday morning-I felt immensely relieved.

When I entered the office the next morning, the Japanese admiral sat there with a cigarette dangling from his mouth. He glared at me as he always did; and said nothing. Six weeks later-thank God-he went back to Tokyo and my worries were ended.

As I have already said, I probably saved my life by sitting down that Sunday afternoon and writing out all the various steps I could take and then writing down the probable consequences of each step and calmly coming to a decision. If I hadn't done that, I might have floundered and hesitated and done the wrong thing on the spur of the moment. If I hadn't thought out my problem and come to a decision, I would have been frantic with worry all Sunday afternoon. I wouldn't have slept that night. I would have gone down to the office Monday morning with a harassed and worried look; and that alone might have aroused the suspicion of the Japanese admiral and spurred him to act.

"Experience has proved to me, time after time, the enormous value of arriving at a decision. It is the failure to arrive at a fixed purpose, the inability to stop going round and round in maddening circles, that drives men to nervous breakdowns and living hells. I find that fifty per cent of my worries vanishes once I arrive at a clear, definite decision; and another forty per cent usually vanishes once I start to carry out that decision.

So I banish about ninety per cent of my worries by taking these four steps:

1. Writing down precisely what I am worrying about.
2. Writing down what I can do about it.
3. Deciding what to do.
4. Starting immediately to carry out that decision.

Galen Litchfield is now the Far Eastern Director for Starr, Park and Freeman, Inc., III John Street, New York, representing large insurance and financial interests.

In fact, as I said before, Galen Litchfield today is one of the most important American businessmen in Asia; and he confesses to me that he owes a large part of his success to this method of analysing worry and meeting it head-on.

Why is his method so superb? Because it is efficient, concrete, and goes directly to the heart of the problem. On top of all that, it is climaxed by the third and indispensable rule: Do something about it. Unless we carry out our action, all our fact-finding and analysis is whistling upwind-it's a sheer waste of energy.

William James said this: "When once a decision is reached and execution is the order of the day, dismiss absolutely all responsibility and care about the outcome." In this case, William James undoubtedly used the word "care" as a synonym for "anxiety". He meant-once you have made a careful decision based on facts, go into action. Don't stop to reconsider. Don't begin to hesitate, worry and retrace your steps. Don't lose yourself in self-doubting which begets other doubts. Don't keep looking back over your shoulder.

I once asked Waite Phillips, one of Oklahoma's most prominent oil men, how he carried out decisions. He replied: "I find that to keep thinking about our problems beyond a certain point is bound to create confusion and worry. There comes a time when any more investigation and thinking are harmful. There comes a time when we must decide and act and never look back."

Why don't you employ Galen Litchfield's technique to one of your worries right now?

Here is:

Question No. 1 -What am I worrying about? (Please pencil the answer to that question in the space below.)

Question No. 2 -What can I do about it? (Please write your answer to that question in the space below.)

Question No. 3 -Here is what I am going to do about it.

Question No. 4 -When am I going to start doing it?

✣ ✣ ✣

Chapter 5

How to Eliminate Fifty Per Cent of Your Business Worries

IF you are a businessman, you are probably saying to yourself right now: "The title of this chapter is ridiculous. I have been running my business for nineteen years; and I certainly know the answers if anybody does. The idea of anybody trying to tell me how I can eliminate fifty per cent of my business worries-it's absurd."

Fair enough-I would have felt exactly the same way myself a few years ago if I had seen this title on a chapter. It promises a lot-and promises are cheap.

Let's be very frank about it: maybe I won't be able to help you eliminate fifty per cent of your business worries. In the last analysis, no one can do that, except yourself. But what I can do is to show you how other people have done it-and leave the rest to you!

You may recall that on page 26 of this book I quoted the world-famous Dr. Alexis Carrel as saying: "Businessmen who do not know how to fight worry die young."

Since worry is that serious, wouldn't you be satisfied if I could help you eliminate even ten per cent of your worries? ... Yes? ... Good! Well, I am going to show you how one business executive eliminated not fifty per cent of his worries, but seventy-five per cent of all the time he formerly spent in conferences, trying to solve business problems.

Furthermore, I am not going to tell you this story about a "Mr. Jones" or a "Mr. X" or "or a man I know in Ohio"- vague stories that you can't check up on. It concerns a very real person-Leon Shimkin, a partner and general manager of one of the foremost publishing houses in the United States: Simon and Schuster, Rockefeller Centre, New York 20, New York.

Here is Leon Shimkin's experience in his own words:

"For fifteen years I spent almost half of every business day holding conferences, discussing problems. Should we do this or that-do nothing at all? We would get tense; twist in our chairs; walk the floor; argue and go around in circles. When night came, I would be utterly exhausted. I fully expected to go on doing this sort of thing for the rest of my life. I had been doing it for fifteen years, and it never occurred to me that there was a better way of doing it. If anyone had told me that I could eliminate three-fourths of all the time I spent in those worried conferences, and three-fourths of my nervous strain-I would have thought he was a wild-eyed, slap-happy, armchair optimist. Yet I devised a plan that did just that. I have been using this plan for eight years. It has performed wonders for my efficiency, my health, and my happiness.

"It sounds like magic-but like all magic tricks, it is extremely simple when you see how it is done.

"Here is the secret: First, I immediately stopped the procedure I had been using in my conferences for fifteen years-a procedure that began with my troubled associates reciting all the details of what had gone wrong, and ending up by asking: 'What shall we do?' Second, I made a new rule-a rule that everyone who wishes to present a problem to me must first prepare and submit a memorandum answering these four questions:

"Question 1: What is the problem?

("In the old days we used to spend an hour or two in a worried conference without anyone's knowing specifically and concretely what the real problem was. We used to work ourselves into a lather discussing our troubles without ever troubling to write out specifically what our problem was.)

"Question 2: What is the cause of the problem?

("As I look back over my career, I am appalled at the wasted hours I have spent in worried conferences without ever trying to find out clearly the conditions which lay at the root of the problem.)

"Question 3: What are all possible solutions of the problem?

("In the old days, one man in the conference would suggest one solution. Someone else would argue with him. Tempers would flare. We would often get clear off the subject, and at the end of the conference no one would have written down all the various things we could do to attack the problem.)

"Question 4: What solution do you suggest?

("I used to go into a conference with a man who had spent hours worrying about a situation and going around in circles without ever once thinking through all possible solutions and then writing down: 'This is the solution I recommend.')

"My associates rarely come to me now with their problems. Why? Because they have discovered that in order to answer these four questions they have to get all the facts and think their problems through. And after they have done that they find, in three-fourths of the cases, they don't have to consult me at all, because the proper solution has popped out like a piece of bread popping out from an electric toaster. Even in those cases where consultation is necessary, the discussion takes about one-third the time formerly required, because it proceeds along an orderly, logical path to a reasoned conclusion.

"Much less time is now consumed in the house of Simon and Schuster in worrying and talking about what is wrong; and a lot more action is obtained toward making those things right."

My friend, Frank Bettger, one of the top insurance men in America, tells me he not only reduced his business worries, but nearly doubled his income, by a similar method.

"Years ago," says Frank Bettger, "when I first started to sell insurance, I was filled with a boundless enthusiasm and love for my work. Then something happened. I became so discouraged that I despised my work and thought of giving it up. I think I would have quit-if I hadn't got the idea, one Saturday morning, of sitting down and trying to get at the root of my worries.

"1. I asked myself first: 'Just what is the problem?.' The problem was: that I was not getting high enough returns for the staggering amount of calls I was making. I seemed to do pretty well at selling a prospect, until the moment came for closing a sale. Then the customer would say: 'Well, I'll think it over, Mr. Bettger. Come and see me again.' It was the time I wasted on these follow-up calls that was causing my depression.

"2. I asked myself: 'What are the possible solutions?' But to get the answer to that one, I had to study the facts. I got out my record book for the last twelve months and studied the figures.

I made an astounding discovery! Right there in black and white, I discovered that seventy per cent of my sales had been closed on the very first interview! Twenty-three per cent of my sales had been closed on the second interview! And only seven per cent of my sales had been closed on those third, fourth, fifth, etc., interviews, which were running me ragged and taking up my time. In other words, I was wasting fully one half of my working day on a part of my business which was responsible for only seven per cent of my sales!

"3. 'What is the answer?' The answer was obvious. I immediately cut out all visits beyond the second interview, and spent the extra time building up new prospects. The results were unbelievable. In a very short time, I had almost doubled the cash value of every visit I made from a call!"

As I said, Frank Bettger is now one of the best-known life-insurance salesmen in America. He is with Fidelity Mutual of Philadelphia, and writes a million dollars' worth of policies a year. But he was on the point of giving up. He was on the point of admitting failure-until analysing the problem gave him a boost on the road to success.

Can you apply these questions to your business problems? To repeat my challenge-they can reduce your worries by fifty per cent. Here they are again:

1. What is the problem?
2. What is the CAUSE of the problem?

3. What are all possible solutions to the problem?
4. What solution do you suggest?

Part Two In A Nutshell

RULE 1 Get the facts. Remember that Dean Hawkes of Columbia University said that " half the worry in the world is caused by people trying to make decisions before they have sufficient knowledge on which to base a decision."

RULE 2 After carefully weighing all the facts, come to a decision.

RULE 3 Once a decision is carefully reached, act! Get busy carrying out your decision-and dismiss all anxiety about the outcome.

RULE 4 When you, or any of your associates are tempted to worry about a problem, write out and answer the following questions:

a. What is the problem?
b. What is the cause of the problem?
c. What are all possible solutions?
d. What is the best solution?

✢ ✢ ✢

PART THREE

How To Break The Worry Habit Before It Breaks You

Chapter 6

How To Crowd Worry Out Of Your Mind

I shall never forget the night, a few years ago, when Marion J. Douglas was a student in one of my classes. (I have not used his real name. He requested me, for personal reasons, not to reveal his identity.) But here is his real story as he told it before one of our adult-education classes. He told us how tragedy had struck at his home, not once, but twice. The first time he had lost his five-year-old daughter, a child he adored. He and his wife thought they couldn't endure that first loss; but, as he said: "Ten months later, God gave us another little girl-and she died in five days."

This double bereavement was almost too much to bear. "I couldn't take it," this father told us. "I couldn't sleep, I couldn't eat, I couldn't rest or relax. My nerves were utterly shaken and my confidence gone." At last he went to doctors; one recommended sleeping pills and another recommended a trip. He tried both, but neither remedy helped. He said: "My body felt as if it were encased in a vice, and the jaws of the vice were being drawn tighter and tighter." The tension of grief-if you have ever been paralysed by sorrow, you know what he meant.

"But thank God, I had one child left-a four-year-old son. He gave me the solution to my problem. One afternoon as I sat around feeling sorry for myself, he asked: 'Daddy, will you build a boat for me?' I was in no mood to build a boat; in fact, I was in no mood to do anything. But my son is a persistent little fellow! I had to give in.

"Building that toy boat took about three hours. By the time it was finished, I realised that those three hours spent building that boat were the first hours of mental relaxation and peace that I had had in months!

"That discovery jarred me out of my lethargy and caused me to do a bit of thinking-the first real thinking I had done in months. I realised that it is difficult to worry while you are busy doing something that requires planning and thinking. In my case, building the boat had knocked worry out of the ring. So I resolved to keep busy.

"The following night, I went from room to room in the house, compiling a list of jobs that ought to be done. Scores of items needed to be repaired: bookcases, stair steps, storm windows, window-shades, knobs, locks, leaky taps. Astonishing as it seems, in the course of two weeks I had made a list of 242 items that needed attention.

"During the last two years I have completed most of them. Besides, I have filled my life with stimulating activities. Two nights per week I attend adult-education classes in New York. I have gone in for civic activities in my home town and I am now chairman of the school board. I attend scores of meetings. I help collect money for the Red Cross and other activities. I am so busy now that I have no time for worry."

No time for worry! That is exactly what Winston Churchill said when he was working eighteen hours a day at the height of the war. When he was asked if he worried about his tremendous responsibilities, he said: "I'm too busy. I have no time for worry."

Charles Kettering was in that same fix when he started out to invent a self-starter for automobiles. Mr. Kettering was, until his recent retirement, vice-president of General Motors in charge of the worldfamous General Motors Research Corporation. But in those days, he was so poor that he had to use the hayloft of a barn as a laboratory. To buy groceries, he had to use fifteen hundred dollars that his wife had made by giving piano lessons; later, he had to borrow five hundred dollars on his life insurance. I asked his wife if she wasn't worried at a time like that. "Yes," she replied, "I was so worried I couldn't sleep; but Mr. Kettering wasn't. He was too absorbed in his work to worry."

The great scientist, Pasteur, spoke of "the peace that is found in libraries and laboratories." Why is peace found there? Because the men in libraries and laboratories are usually too absorbed in their tasks to worry about themselves. Research men rarely have nervous breakdowns. They haven't time for such luxuries.

Why does such a simple thing as keeping busy help to drive out anxiety? Because of a law-one of the most fundamental laws ever revealed by psychology. And that law is: that it is utterly impossible for any human mind, no matter how brilliant, to think of more than one thing at any given time. You don't quite believe it? Very well, then, let's try an experiment.

Suppose you lean right back now, close your eyes, and try, at the same instant, to think of the Statue of Liberty and of what you plan to do tomorrow morning. (Go ahead, try it.)

You found out, didn't you, that you could focus on either thought in turn, but never on both simultaneously? Well, the same thing is true in the field of emotions. We cannot be pepped up and enthusiastic about doing something exciting and feel dragged down by worry at the very same time. One kind of emotion drives out the other. And it was that simple discovery that enabled Army psychiatrists to perform such miracles during the war.

When men came out of battle so shaken by the experience that they were called "psychoneurotic", Army doctors prescribed "Keep 'em busy" as a cure.

Every waking minute of these nerve-shocked men was filled with activity-usually outdoor activity, such as fishing, hunting, playing ball, golf, taking pictures, making gardens, and dancing. They were given no time for brooding over their terrible experiences.

"Occupational therapy" is the term now used by psychiatry when work is prescribed as though it were a medicine. It is not new. The old Greek physicians were advocating it five hundred years before Christ was born!

The Quakers were using it in Philadelphia in Ben Franklin's time. A man who visited a Quaker sanatorium in 1774 was shocked to see that the patients who were mentally ill were busy spinning flax. He thought these poor unfortunates were being exploited-until the Quakers explained that they found that their patients actually improved when they did a little work. It was soothing to the nerves. Any psychiatrist will tell you that work-keeping busy- is one of the best anesthetics ever known for sick nerves. Henry W. Longfellow found that out for himself when he lost his young wife. His wife had been melting some sealing-wax at a candle one day, when her clothes caught on fire. Longfellow heard her cries and tried to reach her in time; but she died from the burns. For a while, Longfellow was so tortured by the memory of that dreadful experience that he nearly went insane; but, fortunately for him, his three small children needed his attention. In spite of his own grief, Longfellow undertook to be father and mother to his children. He took them for walks, told them stories, played games with them, and immortalised their companionship in his poem The Children's Hour. He also translated Dante; and all these duties combined kept him so busy that he forgot himself entirely, and regained his peace of mind. As Tennyson declared when he lost his most intimate friend, Arthur Hallam: "I must lose myself in action, lest I wither in despair."

Most of us have little trouble "losing ourselves in action" while we have our noses to the grindstone and are doing our day's work. But the hours after work-they are the dangerous ones. Just when we're free to enjoy our own leisure, and ought to be happiest-that's when the blue devils of worry attack us. That's when we begin to wonder whether we're getting anywhere in life; whether we're in a rut; whether the boss "meant anything" by that remark he made today; or whether we're getting bald.

When we are not busy, our minds tend to become a near-vacuum. Every student of physics knows that "nature abhors a vacuum". The nearest thing to a vacuum that you and I will probably ever see is the inside of an incandescent electric-light bulb. Break that bulb-and nature forces air in to fill the theoretically empty space.

Nature also rushes in to fill the vacant mind. With what? Usually with emotions. Why? Because emotions of worry, fear, hate, jealousy, and envy are driven by primeval vigour and the dynamic energy of the jungle. Such emotions are so violent that they tend to drive out of our minds all peaceful, happy thoughts and emotions.

James L. Mursell, professor of education, Teachers' College, Columbia, puts it very well when he says: "Worry is most apt to ride you ragged not when you are in action, but when the day's work is done. Your imagination can run riot then and bring up all sorts of ridiculous possibilities and magnify each little blunder. At such a time," he continues, "your mind is like a motor operating without its load. It races and threatens to burn out its bearings or even to tear itself to bits. The remedy for worry is to get completely occupied doing something constructive."

But you don't have to be a college professor to realise this truth and put it into practice. During the war, I met a housewife from Chicago who told me how she discovered for herself that "the remedy for worry is to get completely occupied doing something constructive." I met this woman and her husband in the dining-car while I was travelling from New York to my farm in Missouri. (Sorry I didn't get their names-I never like to give examples without using names and street addresses- details that give authenticity to a story.)

This couple told me that their son had joined the armed forces the day after Pearl Harbour. The woman told me that she had almost wrecked her health worrying over that only son. Where was he? Was he safe? Or in action? Would he be wounded? Killed?

When I asked her how she overcame her worry, she replied: "I got busy." She told me that at first she had dismissed her maid and tried to keep busy by doing all her housework herself. But that didn't help much. "The trouble was," she said, "that I could do my housework almost mechanically, without using my mind. So I kept on worrying. While making the beds and washing the dishes I realised I needed some new kind of work that would keep me busy both mentally and physically every hour of the day. So I took a job as a saleswoman in a large department store.

"That did it," she said. "I immediately found myself in a whirlwind of activity: customers swarming around me, asking for prices, sizes, colours. Never a second to think of anything except my immediate duty; and when night came, I could think of nothing except getting off my aching feet. As soon as I ate dinner, I fell into bed and instantly became unconscious. I had neither the time nor the energy to worry."

She discovered for herself what John Cowper Powys meant when he said, in *The Art of Forgetting the Unpleasant*: "A certain comfortable security, a certain profound inner peace, a kind of happy numbness, soothes the nerves of the human animal when absorbed in its allotted task."

And what a blessing that it is so! Osa Johnson, the world's most famous woman explorer, recently told me how she found release from worry and grief. You may have read the story of her life. It is called *I Married Adventure*. If any woman ever married adventure, she certainly did. Martin Johnson married her when she was sixteen and lifted her feet off the sidewalks of Chanute, Kansas, and set them down on the wild jungle trails of Borneo. For a quarter of a century, this Kansas couple travelled all over the world, making motion pictures of the vanishing wild life of Asia and Africa. Back in America nine years ago, they were on a lecture tour, showing their famous films. They took a plane out of Denver, bound for the Coast. The plane plunged into a mountain. Martin Johnson was killed instantly. The doctors said Osa would never leave her bed again. But they didn't know Osa Johnson. Three months later, she was in a wheel chair, lecturing before large audiences. In fact, she addressed over a hundred audiences that season-all from a wheel chair. When I asked her why she did it, she replied: "I did it so that I would have no time for sorrow and worry."

Osa Johnson had discovered the same truth that Tennyson had sung about a century earlier: "I must lose myself in action, lest I wither in despair."

Admiral Byrd discovered this same truth when he lived all alone for five months in a shack that was literally buried in the great glacial ice-cap that covers the South Pole-an ice-cap that holds nature's oldest secrets-an ice-cap covering an unknown continent larger than the United States and Europe combined. Admiral Byrd spent five months there alone. No other living creature of any kind existed within a hundred miles. The cold was so intense that he could hear his breath freeze and crystallise as the wind blew it past his ears. In his book *Alone*, Admiral Byrd tells all about those five months he spent in bewildering and soul-shattering darkness. The days were as black as the nights. He had to keep busy to preserve his sanity.

"At night," he says, "before blowing out the lantern, I formed the habit of blocking out the morrow's work. It was a case of assigning myself an hour, say, to the Escape Tunnel, half an hour to leveling drift, an hour to straightening up the fuel drums, an hour to cutting bookshelves in the walls of the food tunnel, and two hours to renewing a broken bridge in the man-hauling sledge. ...

"It was wonderful," he says, "to be able to dole out time in this way. It brought me an extraordinary sense of command over myself. ..." And he adds: "Without that or an equivalent, the days would have been without purpose; and without purpose they would have ended, as such days always end, in disintegration."

Note that last again: "Without purpose, the days would have ended, as such days always end, in disintegration."

If you and I are worried, let's remember that we can use good old-fashioned work as a medicine. That was said by no less an authority than the late Dr. Richard C. Cabot, formerly professor of clinical medicine at Harvard. In his book *What Men Live By*, Dr. Cabot says: "As a physician, I have had the happiness of seeing work cure many persons who have suffered from trembling palsy of the soul which results from overmastering doubts, hesitations, vacillation and fear. ... Courage given us by our work is like the self-reliance which Emerson has made forever glorious."

If you and I don't keep busy-if we sit around and brood- we will hatch out a whole flock of what Charles Darwin used to call the "wibber gibbers". And the "wibber gibbers" are nothing but oldfashioned gremlins that will run us hollow and destroy our power of action and our power of will.

I know a businessman in New York who fought the "wibber gibbers" by getting so busy that he had no time to fret and stew. His name is Tremper Longman, and his office is at 40 Wall Street. He was a student in one of my adult-education classes; and his talk on conquering worry was so interesting, so impressive, that I asked him to have supper with me after class; and we sat in a restaurant until long past midnight, discussing his experiences. Here is the story he told me: "Eighteen years ago, I was so worried I had insomnia. I was tense, irritated, and jittery. I felt I was headed for a nervous breakdown.

"I had reason to be worried. I was treasurer of the Crown Fruit and Extract Company, 418 West Broadway, New York. We had half a million dollars invested in strawberries packed in gallon tins. For twenty years, we had been selling these gallon tins of strawberries to manufactures of ice cream. Suddenly our sales stopped because the big ice-cream makers, such as National Dairy and Borden's, were rapidly increasing their production and were saving money and time by buying strawberries packed in barrels.

"Not only were we left with half a million dollars in berries we couldn't sell, but we were also under contract to buy a million dollars more of strawberries in the next twelve months! We had already borrowed $350,000 from the banks. We couldn't possibly pay off or renew these loans. No wonder I was worried!

"I rushed out to Watsonville, California, where our factory was located, and tried to persuade our president that conditions had changed, that we were facing ruin. He refused to believe it. He blamed our New York office for all the trouble-poor salesmanship.

"After days of pleading, I finally persuaded him to stop packing more strawberries and to sell our new supply on the fresh berry market in San Francisco. That almost solved our problems. I should have been able to stop worrying then; but I couldn't. Worry is a habit; and I had that habit.

"When I returned to New York, I began worrying about everything; the cherries we were buying in Italy, the pineapples we were buying in Hawaii, and so on. I was tense, jittery, couldn't sleep; and, as I have already said, I was heading for a nervous breakdown.

"In despair, I adopted a way of life that cured my insomnia and stopped my worries. I got busy. I got so busy with problems demanding all my faculties that I had no time to worry. I had been working seven hours a day. I now began working fifteen and sixteen hours a day. I got down to the office every morning at eight o'clock and stayed there every night until almost midnight. I took on new duties, new responsibilities. When I got home at midnight, I was so exhausted when I fell in bed that I became unconscious in a few seconds.

"I kept up this programme for about three months. I had broken the habit of worry by that time, so I returned to a normal working day of seven or eight hours. This event occurred eighteen years ago. I have never been troubled with insomnia or worry since then."

George Bernard Shaw was right. He summed it all up when he said: "The secret of being miserable is to have the leisure to bother about whether you are happy or not." So don't bother to think about it! Spit on your hands and get busy. Your blood will start circulating; your mind will start ticking -and pretty soon this whole positive upsurge of life in your body will drive worry from your mind. Get busy. Keep busy. It's the cheapest kind of medicine there is on this earth-and one of the best.

To break the worry habit, here is Rule 1:

3.1

RULE 1

Keep busy. The worried person must lose himself in action, lest he wither in despair.

✦ ✦ ✦

Chapter 7

Don't Let the Beetles Get You Down

Here is a dramatic story that I'll probably remember as long as I live. It was told to me by Robert Moore, of 14 Highland Avenue, Maplewood, New Jersey.

"I learned the biggest lesson of my life in March, 1945," he said, "I learned it under 276 feet of water off the coast of Indo-China. I was one of eighty-eight men aboard the submarine *Baya S.S. 31B*. We had discovered by radar that a small Japanese convoy was coming our way. As daybreak approached, we submerged to attack. I saw through the periscope a Jap destroyer escort, a tanker, and a minelayer. We fired three torpedoes at the destroyer escort, but missed. Something went haywire in the mechanics of each torpedo. The destroyer, not knowing that she had been attacked, continued on. We were getting ready to attack the last ship, the minelayer, when suddenly she turned and came directly at us. (A Jap plane had spotted us under sixty feet of water and had radioed our position to the Jap minelayer.) We went down to 150 feet, to avoid detection, and rigged for a depth charge. We put extra bolts on the hatches; and, in order to make our sub absolutely silent, we turned off the fans, the cooling system, and all electrical gear.

"Three minutes later, all hell broke loose. Six depth charges exploded all around us and pushed us down to the ocean floor -a depth of 276 feet. We were terrified. To be attacked in less than a thousand feet of water is dangerous-less than five hundred feet is almost always fatal. And we were being attacked in a trifle more than half of five hundred feet of water -just about knee-deep, as far as safety was concerned. For fifteen hours, that Jap minelayer kept dropping depth charges.

"If a depth charge explodes within seventeen feet of a sub, the concussion will blow a hole in it. Scores of these depth charges exploded within fifty feet of us. We were ordered 'to secure'- to lie quietly in our bunks and remain calm. I was so terrified I could hardly breathe. 'This is death,' I kept saying to myself over and over. 'This is death! ... This is death!' With the fans and cooling system turned off, the air inside the sub was over a hundred degrees; but I was so chilled with fear that I put on a sweater and a fur-lined jacket; and still I trembled with cold. My teeth chattered. I broke out in a cold, clammy sweat. The attack continued for fifteen hours. Then ceased suddenly. Apparently the Jap minelayer had exhausted its supply of depth charges, and steamed away. Those fifteen hours of attack seemed like fifteen million years. All my life passed before me in review.

"I remembered all the bad things I had done, all the little absurd things I had worried about. I had been a bank clerk before I joined the Navy. I had worried about the long hours, the poor pay, the poor prospects of advancement. I had worried because I couldn't own my own home, couldn't buy a new car, couldn't buy my wife nice clothes. How I had hated my old boss, who was always nagging and scolding! I remembered how I would come home at night sore and grouchy and quarrel with my wife over trifles. I had worried about a scar on my forehead-a nasty cut from an auto accident.

"How big all these worries seemed years ago! But how absurd they seemed when depth charges were threatening to blow me to kingdom come. I promised myself then and there that if I ever saw the sun and the stars again, I would never, never worry again. Never! Never! I Never!!! I learned more about the art of living in those fifteen terrible hours in that submarine than I had learned by studying books for four years in Syracuse University."

We often face the major disasters of life bravely-and then let the trifles, the "pains in the neck", get us down. For example, Samuel Pepys tells in his Diary about seeing Sir Harry Vane's head chopped off in London. As Sir Harry mounted the platform, he was not pleading for his life, but was pleading with the executioner not to hit the painful boil on his neck!

That was another thing that Admiral Byrd discovered down in the terrible cold and darkness of the polar nights-that his men fussed more about the ' 'pains in the neck" than about the big things. They bore, without complaining, the dangers, the hardships, and the cold that was often eighty degrees below zero. "But," says Admiral Byrd, "I know of bunkmates who quit speaking because each suspected the other of inching his gear into the other's allotted space; and I knew of one who could not eat unless he could find a place in the mess hall out of sight of the Fletcherist who solemnly chewed his food twenty-eight times before swallowing."

"In a polar camp," says Admiral Byrd, "little things like that have the power to drive even disciplined men to the edge of insanity."

And you might have added, Admiral Byrd, that "little things" in marriage drive people to the edge of insanity and cause "half the heartaches in the world."

At least, that is what the authorities say. For example, Judge Joseph Sabath of Chicago, after acting as arbiter in more than forty thousand unhappy marriages, declared: "Trivialities are at the bottom of most marital unhappiness"; and Frank S. Hogan, District Attorney of New York County, says: "Fully half the cases in our criminal courts originate in little things. Bar-room bravado, domestic wrangling, an insulting remark, a disparaging word, a rude action-those are the little things that lead to assault and murder. Very few of us are cruelly and greatly wronged. It is the small blows to our self-esteem, the indignities, the little jolts to our vanity, which cause half the heartaches in the world."

When Eleanor Roosevelt was first married, she "worried for days" because her new cook had served a poor meal. "But if that happened now," Mrs. Roosevelt says, "I would shrug my shoulders and forget it." Good. That is acting like an adult emotionally. Even Catherine the Great, an absolute autocrat, used to laugh the thing off when the cook spoiled a meal.

Mrs. Carnegie and I had dinner at a friend's house in Chicago. While carving the meat, he did something wrong. I didn't notice it; and I wouldn't have cared even if I had noticed it. But his wife saw it and jumped down his throat right in front of us. "John," she cried, "watch what you are doing! Can't you ever learn to serve properly!"

Then she said to us: "He is always making mistakes. He just doesn't try." Maybe he didn't try to carve; but I certainly give him credit for trying to live with her for twenty years. Frankly, I would rather have eaten a couple of hot dogs with mustard-in an atmosphere of peace-than to have dined on Peking duck and shark fins while listening to her scolding.

Shortly after that experience, Mrs. Carnegie and I had some friends at our home for dinner. Just before they arrived, Mrs. Carnegie found that three of the napkins didn't match the tablecloth.

"I rushed to the cook," she told me later, "and found that the other three napkins had gone to the laundry. The guests were at the door. There was no time to change. I felt like bursting into tears! All I could think was: 'Why did this stupid mistake have to spoil my whole evening?' Then I thought-well, why let it? I went in to dinner, determined to have a good time. And I did. I would much rather our friends think I was a sloppy housekeeper," she told me, "than a nervous, bad-tempered one. And anyhow, as far as I could make out, no one noticed the napkins!"

A well-known legal maxim says: De minimis non curat lex-"the law does not concern itself with trifles." And neither should the worrier-if he wants peace of mind.

Much of the time, all we need to overcome the annoyance of trifles is to affect a shifting of emphasisset up a new, and pleasurable, point of view in the mind. My friend Homer Croy, who wrote *They Had to See Paris* and a dozen other books, gives a wonderful example of how this can be done. He used to be driven half crazy, while working on a book, by the rattling of the radiators in his New York apartment. The steam would bang and sizzle-and he would sizzle with irritation as he sat at his desk.

"Then," says Homer Croy, "I went with some friends on a camping expedition. While listening to the limbs crackling in the roaring fire, I thought how much they sounded like the crackling of the radiators. Why should I like one and hate the other? When I went home I said to myself: 'The crackling of the limbs in the fire was a pleasant sound; the sound of the radiators is about the same-I'll go to sleep and not worry about the noise.' And I did. For a few days I was conscious of the radiators; but soon I forgot all about them.

"And so it is with many petty worries. We dislike them and get into a stew, all because we exaggerate their importance. ..."

Disraeli said: "Life is too short to be little." "Those words," said Andre Maurois in This Week magazine, "have helped me through many a painful experience: often we allow ourselves to be upset by small things we should despise and forget. ... Here we are on this earth, with only a few more decades to live, and we lose many irreplaceable hours brooding over grievances that, in a year's time, will be forgotten by us and by everybody. No, let us devote our life to worth-while actions and feelings, to great thoughts, real affections and enduring undertakings. For life is too short to be little."

Even so illustrious a figure as Rudyard Kipling forgot at times that "Life is too short to be little". The result? He and his brother-in-law fought the most famous court battle in the history of Vermont-a battle so celebrated that a book has been written about it: Rudyard Kipling's Vermont Feud.

The story goes like this: Kipling married a Vermont girl, Caroline Balestier, built a lovely home in Brattleboro, Vermont; settled down and expected to spend the rest of his life there. His brother-in-law, Beatty Balestier, became Kipling's best friend. The two of them worked and played together.

Then Kipling bought some land from Balestier, with the understanding that Balestier would be allowed to cut hay off it each season. One day, Balestier found Kipling laying out a flower garden on this hayfield. His blood boiled. He hit the ceiling. Kipling fired right back. The air over the Green Mountains of Vermont turned blue!

A few days later, when Kipling was out riding his bicycle, his brother-in-law drove a wagon and a team of horses across the road suddenly and forced Kipling to take a spill. And Kipling the man who wrote: "If you can keep your head when all about you are losing theirs and blaming it on you"- he lost his own head, and swore out a warrant for Balestier's arrest. I A sensational trial followed. Reporters from the big cities poured into the town. The news flashed around the world. Nothing was settled. This quarrel caused Kipling and his wife to abandon their American home for the rest of their lives. All that worry and bitterness over a mere trifle! A load of hay.

Pericles said, twenty-four centuries ago: "Come, gentlemen, we sit too long on trifles." We do, indeed!

Here is one of the most interesting stories that Dr. Harry Emerson Fosdick ever told-a story about the battles won and lost by a giant of the forest:

On the slope of Long's Peak in Colorado lies the ruin of a gigantic tree. Naturalists tell us that it stood for some four hundred years. It was a seedling when Columbus landed at San Salvador, and half grown when the Pilgrims settled at Plymouth. During the course of its long life it was struck by lightning fourteen times, and the innumerable avalanches and storms of four centuries thundered past it. It survived them all. In the end, however, an army of beetles attacked the tree and leveled it to the ground. The insects ate their way through the bark and gradually destroyed the inner strength of the tree by their tiny but incessant attacks. A forest giant which age had not withered, nor lightning blasted, nor storms subdued, fell at last before beetles so small that a man could crush them between his forefinger and his thumb.

Aren't we all like that battling giant of the forest? Don't we manage somehow to survive the rare storms and avalanches and lightning blasts of life only to let our hearts be eaten out by little beetles of worry-little beetles that could be crushed between a finger and a thumb?

A few years ago, I travelled through the Teton National Park, in Wyoming, with Charles Seifred, highway superintendent for the state of Wyoming, and some of his friends. We were all going to visit the John D. Rockefeller estate in the park. But the car in which I was riding took the wrong turn, got lost, and drove up to the entrance of the estate an hour after the other cars had gone in. Mr. Seifred had the key that unlocked the private gate, so he waited in the hot, mosquito-infested woods for an hour until we arrived. The mosquitoes were enough to drive a saint insane. But they couldn't triumph over Charles Seifred. While waiting for us, he cut a limb off an aspen tree-and made a whistle of it. When we arrived, was he cussing the mosquitoes? No, he was playing his whistle. I have kept that whistle as a memento of a man who knew how to put trifles in their place.

To break the worry habit before it breaks you, here is Rule 2:

3.2

RULE 2

Let's not allow ourselves to be upset by small things we should despise and forget. Remember "Life is too short to be little."

✢ ✢ ✢

Chapter 8

A Law That Will Outlaw Many of Your Worries

As a child, I grew up on a Missouri farm; and one day, while helping my mother pit cherries, I began to cry. My mother said: "Dale, what in the world are you crying about?" I blubbered: "I'm afraid I am going to be buried alive!"

I was full of worries in those days. When thunderstorms came, I worried for fear I would be killed by lightning. When hard times came, I worried for fear we wouldn't have enough to eat. I worried for fear I would go to hell when I died. I was terrified for fear an older boy, Sam White, would cut off my big ears-as he threatened to do. I worried for fear girls would laugh at me if I tipped my hat to them. I worried for fear no girl would ever be willing to marry me. I worried about what I would say to my wife immediately after we were married. I imagined that we would be married in some country church, and then get in a surrey with fringe on the top and ride back to the farm ... but how would I be able to keep the conversation going on that ride back to the farm? How? How? I pondered over that earth-shaking problem for many an hour as I walked behind the plow.

As the years went by, I gradually discovered that ninety-nine per cent of the things I worried about never happened.

For example, as I have already said, I was once terrified of lightning; but I now know that the chances of my being killed by lightning in any one year are, according to the National Safety Council, only one in three hundred and fifty thousand.

My fear of being buried alive was even more absurd: I don't imagine that one person in ten million is buried alive; yet I once cried for fear of it.

One person out of every eight dies of cancer. If I had wanted something to worry about, I should have worried about cancer -instead of being killed by lightning or being buried alive.

To be sure, I have been talking about the worries of youth and adolescence. But many of our adult worries are almost as absurd. You and I could probably eliminate nine-tenths of our worries right now if we would cease our fretting long enough to discover whether, by the law of averages, there was any real justification for our worries.

The most famous insurance company on earth-Lloyd's of London-has made countless millions out of the tendency of everybody to worry about things that rarely happen. Lloyd's of London bets people that the disasters they are worrying about will never occur. However, they don't call it betting. They call it insurance. But it is really betting based on the law of averages. This great insurance firm has been going strong for two hundred years; and unless human nature changes, it will still be going strong fifty centuries from now by insuring shoes and ships and sealing-wax against disasters that, by the law of average, don't happen nearly so often as people imagine.

If we examine the law of averages, we will often be astounded at the facts we uncover. For example, if I knew that during the next five years I would have to fight in a battle as bloody as the Battle of Gettysburg, I would be terrified. I would take out all the life insurance I could get. I would draw up my will and set all my earthly affairs in order. I would say: "I'll probably never live through that battle, so I had better make the most of the few years I have left." Yet the facts are that, according to the law of averages, it is just as dangerous, just as fatal, to try to live from age fifty to age fifty-five in peacetime as it was to fight in the Battle of Gettysburg. What I am trying to say is this: in times of peace, just as many people die per thousand between the ages of fifty and fifty-five as were killed per thousand among the 163,000 soldiers who fought at Gettysburg.

I wrote several chapters of this book at James Simpson's Num-Ti-Gah Lodge, on the shore of Bow Lake in the Canadian Rockies. While stopping there one summer, I met Mr. and Mrs. Herbert H. Salinger, of 2298 Pacific Avenue, San

Francisco. Mrs. Salinger, a poised, serene woman, gave me the impression that she had never worried. One evening in front of the roaring fireplace, I asked her if she had ever been troubled by worry. "Troubled by it?" she said. "My life was almost ruined by it. Before I learned to conquer worry, I lived through eleven years of self-made hell. I was irritable and hottempered. I lived under terrific tension. I would take the bus every week from my home in San Mateo to shop in San Francisco. But even while shopping, I worried myself into a dither: maybe I had left the electric iron connected on the ironing board. Maybe the house had caught fire. Maybe the maid had run off and left the children. Maybe they had been out on their bicycles and been killed by a car. In the midst of my shopping, I would often worry myself into a cold perspiration and rush out and take the bus home to see if everything was all right. No wonder my first marriage ended in disaster.

"My second husband is a lawyer-a quiet, analytical man who never worries about anything. When I became tense and anxious, he would say to me: 'Relax. Let's think this out. ....What are you really worrying about? Let's examine the law of averages and see whether or not it is likely to happen.'

"For example, I remember the time we were driving from Albuquerque, New Mexico, to the Carlsbad Caverns-driving on a dirt road-when we were caught in a terrible rainstorm.

"The car was slithering and sliding. We couldn't control it. I was positive we would slide off into one of the ditches that flanked the road; but my husband kept repeating to me: 'I am driving very slowly. Nothing serious is likely to happen. Even if the car does slide into the ditch, by the law of averages, we won't be hurt.' His calmness and confidence quieted me.

"One summer we were on a camping trip in the Touquin Valley of the Canadian Rockies. One night we were camping seven thousand feet above sea level, when a storm threatened to tear our tents to shreds. The tents were tied with guy ropes to a wooden platform. The outer tent shook and trembled and screamed and shrieked in the wind. I expected every minute to see our tent torn loose and hurled through the sky. I was terrified! But my husband kept saying: 'Look, my dear, we are travelling with Brewster's guides. Brewster's know what they are doing. They have been pitching tents in these mountains for sixty years. This tent has been here for many seasons. It hasn't blown down yet and, by the law of averages, it won't blow away tonight; and even if it does, we can take shelter in another tent. So relax.' I did; and I slept soundly the balance of the night.

"A few years ago an infantile-paralysis epidemic swept over our part of California. In the old days, I would have been hysterical. But my husband persuaded me to act calmly. We took all the precautions we could; we kept our children away from crowds, away from school and the movies. By consulting the Board of Health, we found out that even during the worst infantile-paralysis epidemic that California had ever known up to that time, only 1,835 children had been stricken in the entire state of California. And that the usual number was around two hundred or three hundred. Tragic as those figures are, we nevertheless felt that, according to the law of averages, the chances of any one child being stricken were remote."

'By the law of averages, it won't happen.' That phrase has destroyed ninety per cent of my worries; and it has made the past twenty years of my life beautiful and peaceful beyond my highest expectations."

General George Crook-probably the greatest Indian fighter in American history-says in his Autobiography that "nearly all the worries and unhappiness" of the Indians "came from their imagination, and not from reality."

As I look back across the decades, I can see that that is where most of my worries came from also. Jim Grant told me that that had been his experience, too. He owns the James A. Grant Distributing Company, 204 Franklin Street, New York City. He orders from ten to fifteen car-loads of Florida oranges and grapefruit at a time. He told me that he used to torture himself with such thoughts as: What if there's a train wreck? What if my fruit is strewn all over the countryside? What if a bridge collapses as my cars are going across it? Of course, the fruit was insured; but he feared that if he didn't deliver his fruit on time, he might risk the loss of his market. He worried so much that he feared he had stomach ulcers and went to a doctor. The doctor told him there was nothing wrong with him except jumpy nerves. "I saw the light then," he said, "and began to ask myself questions. I said to myself: 'Look here, Jim Grant, how many fruit cars have you handled over the years?' The answer was: 'About twenty-five thousand.' Then I asked myself: 'How many of those cars were ever wrecked?' The answer was: 'Oh-maybe five.' Then I said to myself: 'Only five-out of twenty-five thousand? Do you know what that means? A ratio of five thousand to one! In other words, by the law of averages, based on experience, the chances are five thousand to one against one of your cars ever being wrecked. So what are you worried about?'

"Then I said to myself: 'Well, a bridge may collapse!' Then I asked myself: 'How many cars have you actually lost from a bridge collapsing?' The answer was-'None.' Then I said to myself: 'Aren't you a fool to be worrying yourself into stomach ulcers over a bridge which has never yet collapsed, and over a railroad wreck when the chances are five thousand to one against it!'

"When I looked at it that way," Jim Grant told me, "I felt pretty silly. I decided then and there to let the law of averages do the worrying for me-and I have not been troubled with my 'stomach ulcer' since!"

When Al Smith was Governor of New York, I heard him answer the attacks of his political enemies by saying over and over: "Let's examine the record ... let's examine the record." Then he proceeded to give the facts. The next time you and I are worrying about what may happen, let's take a tip from wise old Al Smith: let's examine the record and see what basis there is, if any, for our gnawing anxieties. That is precisely what Frederick J. Mahlstedt did when he feared he was lying in his grave. Here is his story as he told it to one of our adult-education classes in New York:

"Early in June, 1944, I was lying in a slit trench near Omaha Beach. I was with the 999th Signal Service Company, and we had just 'dug in' in Normandy. As I looked around at that slit trench-just a rectangular hole in the ground-I said to myself: 'This looks just like a grave.' When I lay down and tried to sleep in it, it felt like a grave. I couldn't help saying to myself: 'Maybe this is my grave.' When the German bombers began coming over at 11 p.m., and the bombs started falling, I was scared stiff. For the first two or three nights I couldn't sleep at all. By the fourth or fifth night, I was almost a nervous wreck. I knew that if I didn't do something, I would go stark crazy. So I reminded myself that five nights had passed, and I was still alive; and so was every man in our outfit. Only two had been injured, and they had been hurt, not by German bombs, but by falling flak, from our own anti-aircraft guns. I decided to stop worrying by doing something constructive. So I built a thick wooden roof over my slit trench, to protect myself from flak. I thought of the vast area over which my unit was spread. I told myself that the only way I could be killed in that deep, narrow slit trench was by a direct hit; and I figured out that the chance of a direct hit on me was not one in ten thousand. After a couple of nights of looking at it in this way, I calmed down and slept even through the bomb raids!"

The United States Navy used the statistics of the law of averages to buck up the morale of their men. One ex-sailor told me that when he and his shipmates were assigned to high-octane tankers, they were worried stiff. They all believed that if a tanker loaded with high-octane gasoline was hit by a torpedo, it exploded and blew everybody to kingdom come.

But the U.S. Navy knew otherwise; so the Navy issued exact figures, showing that out of one hundred tankers hit by torpedoes sixty stayed afloat; and of the forty that did sink, only five sank in less than ten minutes. That meant time to get off the ship-it also meant casualties were exceedingly small. Did this help morale? "This knowledge of the law of averages wiped out my jitters," said Clyde W. Maas, of 1969 Walnut Street, St. Paul, Minnesota-the man who told this story. "The whole crew felt better. We knew we had a chance; and that, by the law of averages, we probably wouldn't be killed." To break the worry habit before it breaks you-here is Rule 3:

3.3

RULE 3

"Let's examine the record."
Let's ask ourselves: "What are the chances, according to the law of averages, that this event I am worrying about will ever occur?"

✦ ✦ ✦

Chapter 9

Co-Operate With The Inevitable

When I was a little boy, I was playing with some of my friends in the attic of an old, abandoned log house in north-west Missouri. As I climbed down out of the attic, I rested my feet on a window-sill for a moment-and then jumped. I had a ring on my left forefinger; and as I jumped, the ring caught on a nail head and tore off my finger.

I screamed. I was terrified. I was positive I was going to die. But after the hand healed, I never worried about it for one split second. What would have been the use? ... I accepted the inevitable.

Now I often go for a month at a time without even thinking about the fact that I have only three fingers and a thumb on my left hand.

A few years ago, I met a man who was running a freight elevator in one of the downtown office buildings in New York. I noticed that his left hand had been cut off at the wrist. I asked him if the loss of that hand bothered him. He said: "Oh, no, I hardly ever think about it. I am not married; and the only time I ever think about it is when I try to thread a needle."

It is astonishing how quickly we can accept almost any situation-if we have to-and adjust ourselves to it and forget about it.

I often think of an inscription on the ruins of a fifteenth-century cathedral in Amsterdam, Holland. This inscription says in Flemish: "It is so. It cannot be otherwise."

As you and I march across the decades of time, we are going to meet a lot of unpleasant situations that are so. They cannot be otherwise. We have our choice. We can either accept them as inevitable and adjust ourselves to them, or we can ruin our lives with rebellion and maybe end up with a nervous breakdown.

Here is a bit of sage advice from one of my favourite philosophers, William James. "Be willing to have it so," he said. "Acceptance of what has happened is the first step to overcoming the consequence of any misfortune." Elizabeth Connley, of 2840 NE 49th Avenue, Portland, Oregon, had to find that out the hard way. Here is a letter that she wrote me recently: "On the very day that America was celebrating the victory of our armed forces in North Africa," the letter says, "I received a telegram from the War Department: my nephew- the person I loved most-was missing in action. A short time later, another telegram arrived saying he was dead.

"I was prostrate with grief. Up to that time, I had felt that life had been very good to me. I had a job I loved. I had helped to raise this nephew. He represented to me all that was fine and good in young manhood. I had felt that all the bread I had cast upon the waters was coming back to me as cake! ... Then came this telegram. My whole world collapsed. I felt there was nothing left to live for. I neglected my work; neglected my friends. I let everything go. I was bitter and resentful. Why did my loving nephew have to be taken? Why did this good boy-with life all before him-why did he have to be killed? I couldn't accept it. My grief was so overwhelming that I decided to give up my work, and go away and hide myself in my tears and bitterness.

"I was clearing out my desk, getting ready to quit, when I came across a letter that I had forgotten-a letter from this nephew who had been killed, a letter he had written to me when my mother had died a few years ago. 'Of course, we will miss her,' the letter said, 'and especially you. But I know you'll carry on. Your own personal philosophy will make you do that. I shall never forget the beautiful truths you taught me. Wherever I am, or how far apart we may be, I shall always remember that you taught me to smile, and to take whatever comes, like a man.'

"I read and reread that letter. It seemed as if he was there beside me, speaking to me. He seemed to be saying to me: 'Why don't you do what you taught me to do? Carry on, no matter what happens. Hide your private sorrows under a smile and carry on.'

"So, I went back to my work. I stopped being bitter and rebellious. I kept saying to myself: 'It is done. I can't change it. But I can and will carry on as he wished me to do.' I threw all my mind and strength into my work. I wrote letters to soldiers-to other people's boys. I joined an adult-education class at night-seeking out new interests and making new friends. I can hardly believe the change that has come over me. I have ceased mourning over the past that is forever gone. I am living each day now with joy, just as my nephew would have wanted me to do. I have made peace with life. I have accepted my fate. I am now living a fuller and more complete life than I had ever known."

Elizabeth Connley, out in Portland, Oregon, learned what all of us will have to learn sooner or later: namely, that we must accept and co-operate with the inevitable. "It is so. It cannot be otherwise." That is not an easy lesson to learn. Even kings on their thrones have to keep reminding themselves of it. The late George V had these framed words hanging on the wall of his library in Buckingham Palace: "Teach me neither to cry for the moon nor over spilt milk." The same thought is expressed by Schopenhauer in this way: "A good supply of resignation is of the first importance in providing for the journey of life."

Obviously, circumstances alone do not make us happy or unhappy. It is the way we react to circumstances that determines our feelings. Jesus said that the kingdom of heaven is within you. That is where the kingdom of hell is, too.

We can all endure disaster and tragedy and triumph over them-if we have to. We may not think we can, but we have surprisingly strong inner resources that will see us through if we will only make use of them. We are stronger than we think.

The late Booth Tarkington always said: "I could take anything that life could force upon me except one thing: blindness. I could never endure that."

Then one day, when he was along in his sixties, Tarkington glanced down at the carpet on the floor. The colours were blurred. He couldn't see the pattern. He went to a specialist. He learned the tragic truth: he was losing his sight. One eye was nearly blind; the other would follow. That which he feared most had come upon him.

And how did Tarkington react to this "worst of all disasters"? Did he feel: "This is it! This is the end of my life"? No, to his amazement, he felt quite gay. He even called upon his humour. Floating "specks" annoyed him; they would swim across his eyes and cut off his vision. Yet when the largest of these specks would swim across his sight, he would say: "Hello! There's Grandfather again! Wonder where he's going on this fine morning!"

How could fate ever conquer a spirit like that? The answer is it couldn't. When total blindness closed in, Tarkington said: "I found I could take the loss of my eyesight, just as a man can take anything else. If I lost all five of my senses, I know I could live on inside my mind. For it is in the mind we see, and in the mind we live, whether we know it or not."

In the hope of restoring his eyesight, Tarkington had to go through more than twelve operations within one year. With local anaesthetic! Did he rail against this? He knew it had to be done. He knew he couldn't escape it, so the only way to lessen his suffering was to take it with grace. He refused a private room at the hospital and went into a ward, where he could be with other people who had troubles, too. He tried to cheer them up. And when he had to submit to repeated operations-fully conscious of what was being done to his eyes-he tried to remember how fortunate he was. "How wonderful!" he said. "How wonderful, that science now has the skill to operate on anything so delicate as the human eye!"

The average man would have been a nervous wreck if he had had to endure more than twelve operations and blindness. Yet Tarkington said: "I would not exchange this experience for a happier one." It taught him acceptance. It taught him that nothing life could bring him was beyond his strength to endure. It taught him, as John Milton discovered, that "It is not miserable to be blind, it is only miserable not to be able to endure blindness."

Margaret Fuller, the famous New England feminist, once offered as her credo: "I accept the Universe!"

When grouchy old Thomas Carlyle heard that in England, he snorted: "By god! she'd better!" Yes, and by god, you and I had better accept the inevitable too.

If we rail and kick against it and grow bitter, we won't change the inevitable; but we will change ourselves. I know. I have tried it.

I once refused to accept an inevitable situation with which I was confronted. I played the fool and railed against it, and rebelled. I turned my nights into hells of insomnia. I brought upon myself everything I didn't want. Finally, after a year of self-torture, I had to accept what I knew from the outset I couldn't possibly alter.

I should have cried out years ago with old Walt Whitman:

Oh, to confront night, storms, hunger,
Ridicule, accident, rebuffs as the trees
and animals do.

I spent twelve years working with cattle; yet I never saw a Jersey cow running a temperature because the pasture was burning up from a lack of rain or because of sleet and cold or because her boy friend was paying too much attention to another heifer. The animals confront night, storms, and hunger calmly; so they never have nervous breakdowns or stomach ulcers; and they never go insane.

Am I advocating that we simply bow down to all the adversities that come our way? Not by a long shot! That is mere fatalism. As long as there is a chance that we can save a situation, let's fight! But when common sense tells us that we are up against something that is so-and cannot be otherwise- then, in the name of our sanity, let's not look before and after and pine for what is not.

The late Dean Hawkes of Columbia University told me that he had taken a Mother Goose rhyme as one of his mottoes:

For every ailment under the sun.
There is a remedy, or there is none;
If there be one, try to find it;
If there be none, never mind it.

While writing this book, I interviewed a number of the leading businessmen of America; and I was impressed by the fact that they co-operated with the inevitable and led lives singularly free from worry. If they hadn't done that, they would have cracked under the strain. Here are a few examples of what I mean:

J.C. Penney, founder of the nation-wide chain of Penney stores, said to me: "I wouldn't worry if I lost every cent I have because I don't see what is to be gained by worrying. I do the best job I possibly can; and leave the results in the laps of the gods."

Henry Ford told me much the same thing. "When I can't handle events," he said, "I let them handle themselves."

When I asked K.T. Keller, president of the Chrysler Corporation, how he kept from worrying, he said: "When I am up against a tough situation, if I can do anything about it, I do it. If I can't, I just forget it. I never worry about the future, because I know no man living can possibly figure out what is going to happen in the future. There are so many forces that will affect that future! Nobody can tell what prompts those forces-or understand them. So why worry about them?" K. T. Keller would be embarrassed if you told him he is a philosopher. He is just a good businessman, yet he has stumbled on the same philosophy that Epictetus taught in Rome nineteen centuries ago. "There is only one way to happiness," Epictetus taught the Romans, "and that is to cease worrying about things which are beyond the power of our will."

Sarah Bernhardt, the "divine Sarah" was an illustrious example of a woman who knew how to cooperate with the inevitable. For half a century, she had been the reigning queen of the theatre on four continents-the best-loved actress on earth. Then when she was seventy-one and broke-she had lost all her money-her physician, Professor Pozzi of Paris, told her he would have to amputate her leg. While crossing the Atlantic, she had fallen on deck during a storm, and injured her leg severely. Phlebitis developed. Her leg shrank. The pain became so intense that the doctor felt her leg had to be amputated. He was almost afraid to tell the stormy, tempestuous "divine Sarah" what had to be done. He fully expected that the terrible news would set off an explosion of hysteria. But he was wrong. Sarah looked at him a moment, and then said quietly: "If it has to be, it has to be." It was fate.

As she was being wheeled away to the operating room, her son stood weeping. She waved to him with a gay gesture and said cheerfully: "Don't go away. I'll be right back."

On the way to the operating room she recited a scene from one of her plays. Someone asked her if she were doing this to cheer herself up. She said: "No, to cheer up the doctors and nurses. It will be a strain on them."

After recovering from the operation, Sarah Bernhardt went on touring the world and enchanting audiences for another seven years.

"When we stop fighting the inevitable," said Elsie MacCormick in a Reader's Digest article, "we release energy which enables us to create a richer life."

No one living has enough emotion and vigour to fight the inevitable and, at the same time, enough left over to create a new life. Choose one or the other. You can either bend with the inevitable sleet-storms of life-or you can resist them and break!

I saw that happen on a farm I own in Missouri. I planted a score of trees on that farm. At first, they grew with astonishing rapidity. Then a sleet-storm encrusted each twig and branch with a heavy coating of ice. Instead of bowing gracefully to their burden, these trees proudly resisted and broke and split under the load-and had to be destroyed. They hadn't learned the wisdom of the forests of the north. I have travelled hundreds of miles through the evergreen forests of Canada, yet I have never seen a spruce or a pine broken by sleet or ice. These evergreen forests know how to bend, how to bow down their branches, how to co-operate with the inevitable.

The masters of jujitsu teach their pupils to "bend like the willow; don't resist like the oak."

Why do you think your automobile tyres stand up on the road and take so much punishment? At first, the manufacturers tried to make a tyre that would resist the shocks of the road. It was soon cut to ribbons. Then they made a tyre that would absorb the shocks of the road. That tyre could "take it". You and I will last longer, and enjoy smoother riding, if we learn to absorb the shocks and jolts along the rocky road of life.

What will happen to you and me if we resist the shocks of life instead of absorbing them? What will happen if we refuse to "bend like the willow" and insist on resisting like the oak? The answer is easy. We will set up a series of inner conflicts. We will be worried, tense, strained, and neurotic.

If we go still further and reject the harsh world of reality and retreat into a dream world of our own making, we will then be insane.

During the war, millions of frightened soldiers had either to accept the inevitable or break under the strain. To illustrate, let's take the case of William H. Casselius, 7126 76th Street, Glendale, New York. Here is a prize-winning talk he gave before one of my adult-education classes in New York:

"Shortly after I joined the Coast Guard, I was assigned to one of the hottest spots on this side of the Atlantic. I was made a supervisor of explosives. Imagine it. Me! A biscuit salesman becoming a supervisor of explosives! The very thought of finding yourself standing on top of thousands of tons of T.N.T. is enough to chill the marrow in a cracker salesman's bones. I was given only two days of instruction; and what I learned filled me with even more terror. I'll never forget my first assignment. On a dark, cold, foggy day, I was given my orders on the open pier of Caven Point, Bayonne, New Jersey.

"I was assigned to Hold No. 5 on my ship. I had to work down in that hold with five longshoremen. They had strong backs, but they knew nothing whatever about explosives. And they were loading blockbusters, each one of which contained a ton of T.N.T.-enough explosive to blow that old ship to kingdom come. These blockbusters were being lowered by two cables. I kept saying to myself: Suppose one of those cables slipped-or broke! Oh, boy! Was I scared! I trembled. My mouth was dry. My knees sagged. My heart pounded. But I couldn't run away. That would be desertion. I would be disgraced-my parents would be disgraced-and I might be shot for desertion. I couldn't run. I had to stay. I kept looking at the careless way those longshoremen were handling those blockbusters. The ship might blow up any minute. After an hour or more of this spine-chilling terror, I began to use a little common sense. I gave myself a good talking to. I said: 'Look here! So you are blown up. So what! You will never know the difference! It will be an easy way to die. Much better than dying by cancer. Don't be a fool. You can't expect to live forever! You've got to do this job-or be shot. So you might as well like it."

"I talked to myself like that for hours; and I began to feel at ease. Finally, I overcame my worry and fears by forcing myself to accept an inevitable situation.

"I'll never forget that lesson. Every time I am tempted now to worry about something I can't possibly change, I shrug my shoulders and say: 'Forget it.' I find that it works-even for a biscuit salesman." Hooray! Let's give three cheers and one cheer more for the biscuit salesman of the Pinafore.

Outside the crucifixion of Jesus, the most famous death scene in all history was the death of Socrates. Ten thousand centuries from now, men will still be reading and cherishing Plato's immortal description of it-one of the most moving and beautiful passages in all literature. Certain men of Athensjealous and envious of old barefooted Socrates-trumped up charges against him and had him tried and condemned to death. When the friendly jailer gave Socrates the poison cup to drink, the jailer said: "Try to bear lightly what needs must be." Socrates did. He faced death with a calmness and resignation that touched the hem of divinity.

"Try to bear lightly what needs must be." Those words were spoken 399 years before Christ was born; but this worrying old world needs those words today more than ever before: "Try to bear lightly what needs must be."

During the past eight years, I have been reading practically every book and magazine article I could find that dealt even remotely with banishing worry. ... Would you like to know what is the best single bit of advice about worry that I have ever discovered in all that reading? Well, here it is-summed up in twenty-seven words-words that you and I ought to paste on our bathroom mirrors, so that each time we wash our faces we could also wash away all worry from our minds. This priceless prayer was written by Dr. Reinhold Niebuhr, Professor of Applied Christianity, Union Theological Seminary, Broadway and 120th Street, New York.

> *God grant me the serenity*
> *To accept the things I cannot change;*
> *The courage to change the things I can;*
> *And the wisdom to know the difference.*

To break the worry habit before it breaks you, Rule 4 is:

3.4

RULE 4

Co-operate with the inevitable.

✦ ✦ ✦

Chapter 10

Put A "Stop-Loss" Order On Your Worries

Would you like to know how to make money on the Stock Exchange? Well, so would a million other people-and if I knew the answer, this book would sell for a fabulous price. However, there's one good idea that some successful operators use. This story was told to me by Charles Roberts, an investment counselor with offices at 17 East 42nd Street, New York.

"I originally came up to New York from Texas with twenty thousand dollars which my friends had given me to invest in the stock market," Charles Roberts told me. "I thought," he continued, "that I knew the ropes in the stock market; but I lost every cent. True, I made a lot of profit on some deals; but I ended up by losing everything.

"I did not mind so much losing my own money," Mr. Roberts explained, "but I felt terrible about having lost my friends' money, even though they could well afford it. I dreaded facing them again after our venture had turned out so unfortunately, but, to my astonishment, they not only were good sports about it, but proved to be incurable optimists.

"I knew I had been trading on a hit-or-miss basis and depending largely on luck and other people's opinions. As H. I. Phillips said, I had been 'playing the stock market by ear'.

"I began to think over my mistakes and I determined that before I went back into the market again, I would try to find out what it was all about. So I sought out and became acquainted with one of the most successful speculators who ever lived: Burton S. Castles. I believed I could learn a great deal from him because he had long enjoyed the reputation of being successful year after year and I knew that such a career was not the result of mere chance or luck.

"He asked me a few questions about how I had traded before and then told me what I believe is the most important principle in trading. He said: 'I put a stop-loss order on every market commitment I make. If I buy a stock at, say, fifty dollars a share, I immediately place a stop-loss order on it at fortyfive.' That means that when and if the stock should decline as much as five points below its cost, it would be sold automatically, thereby, limiting the loss to five points."

'If your commitments are intelligently made in the first place,' the old master continued, 'your profits will average ten, twenty-five, or even fifty points. Consequently, by limiting your losses to five points, you can be wrong more than half of the time and still make plenty of money?'

"I adopted that principle immediately and have used it ever since. It has saved my clients and me many thousands of dollars."

"After a while I realised that the stop-loss principle could be used in other ways besides in the stock market. I began to place a stop-loss order on any and every kind of annoyance and resentment that came to me. It has worked like magic.

"For example, I often have a luncheon date with a friend who is rarely on time. In the old days, he used to keep me stewing around for half my lunch hour before he showed up. Finally, I told him about my stop-loss orders on my worries. I said: 'Bill, my stop-loss order on waiting for you is exactly ten minutes. If you arrive more than ten minutes late, our luncheon engagement will be sold down the river-and I'll be gone.'"

Man alive! How I wish I had had the sense, years ago, to put stop-loss orders on my impatience, on my temper, on my desire for self-justification, on my regrets, and on all my mental and emotional strains. Why didn't I have the horse sense to size up each situation that threatened to destroy my peace of mind and say to myself: "See here, Dale Carnegie, this situation is worth just so much fussing about and no more?" ... Why didn't I?

However, I must give myself credit for a little sense on one occasion, at least. And it was a serious occasion, too-a crisis in my life-a crisis when I stood watching my dreams and my plans for the future and the work of years vanish into thin air. It happened like this. In my early thirties, I had decided to spend my life writing novels. I was going to be a second Frank Norris or Jack London or Thomas Hardy. I was so in earnest that I spent two years in Europe - where I would live cheaply with dollars during the period of wild, printing-press money that followed the First World War. I spent two years there, writing my magnum opus. I called it The Blizzard.

The title was a natural, for the reception it got among publishers was as cold as any blizzard that ever howled across the plains of the Dakotas. When my literary agent told me it was worthless, that I had no gift, no talent, for fiction, my heart almost stopped. I left his office in a daze. I couldn't have been more stunned if he had hit me across the head with a club. I was stupefied. I realised that I was standing at the crossroads of life, and had to make a tremendous decision. What should I do? Which way should I turn? Weeks passed before I came out of the daze. At that time, I had never heard of the phrase "put a stop-loss order on your worries". But as I look back now, I can see that I did just that. I wrote off my two years of sweating over that novel for just what they were worth - a noble experiment - and went forward from there. I returned to my work of organising and teaching adult-education classes, and wrote biographies in my spare time - biographies and non-fiction books such as the one you are reading now.

Am I glad now that I made that decision? Glad? Every time I think about it now I feel like dancing in the street for sheer joy! I can honestly say that I have never spent a day or an hour since, lamenting the fact that I am not another Thomas Hardy.

One night a century ago, when a screech owl was screeching in the woods along the shore of Walden Pond, Henry Thoreau dipped his goose quill into his homemade ink and wrote in his diary: "The cost of a thing is the amount of what I call life, which is required to be exchanged for it immediately or in the long run."

To put it another way: we are fools when we overpay for a thing in terms of what it takes out of our very existence.

Yet that is precisely what Gilbert and Sullivan did. They knew how to create gay words and gay music, but they knew distressingly little about how to create gaiety in their own lives. They created some of the loveliest light operas that ever delighted the world: Patience, Pinafore, The Mikado. But they couldn't control their tempers. They embittered their years over nothing more than the price of a carpet! Sullivan ordered a new carpet for the theatre they had bought. When Gilbert saw the bill, he hit the roof. They battled it out in court, and never spoke to one another again as long as they lived. When Sullivan wrote the music for a new production, he mailed it to Gilbert; and when Gilbert wrote the words, he mailed it back to Sullivan. Once they had to take a curtain call together, but they stood on opposite sides of the stage and bowed in different directions, so they wouldn't see one another. They hadn't the sense to put a stop-loss order on their resentments, as Lincoln did.

Once, during the Civil War, when some of Lincoln's friends were denouncing his bitter enemies, Lincoln said: "You have more of a feeling of personal resentment than I have. Perhaps I have too little of it; but I never thought it paid. A man doesn't have the time to spend half his life in quarrels. If any man ceases to attack me, I never remember the past against him."

I wish an old aunt of mine-Aunt Edith-had had Lincoln's forgiving spirit. She and Uncle Frank lived on a mortgaged farm that was infested with cockleburs and cursed with poor soil and ditches. They had tough going-had to squeeze every nickel. But Aunt Edith loved to buy a few curtains and other items to brighten up their bare home. She bought these small luxuries on credit at Dan Eversole's drygoods store in Maryville, Missouri. Uncle Frank worried about their debts. He had a farmer's horror of running up bills, so he secretly told Dan Eversole to stop letting his wife buy on credit. When she heard that, she hit the roof-and she was still hitting the roof about it almost fifty years after it had happened. I have heard her tell the story-not once, but many times. The last time I ever saw her, she was in her late seventies. I said to her; "Aunt Edith, Uncle Frank did wrong to humiliate you; but don't you honestly feel that your complaining about it almost half a century after it happened is infinitely worse than what he did?" (I might as well have said it to the moon.)

Aunt Edith paid dearly for the grudge and bitter memories that she nourished. She paid for them with her own peace of mind.

When Benjamin Franklin was seven years old, he made a mistake that he remembered for seventy years. When he was a lad of seven, he fell in love with a whistle. He was so excited about it that he went into the toyshop, piled all his coppers on the counter, and demanded the whistle without even asking its price. "I then came home," he wrote to a friend seventy years later, "and went whistling all over the house, much pleased with my whistle." But when his older brothers and sisters found out that he had paid far more for his whistle than he should have paid, they gave him the horse laugh; and, as he said: "I cried with vexation."

Years later, when Franklin was a world-famous figure, and Ambassador to France, he still remembered that the fact that he had paid too much for his whistle had caused him "more chagrin than the whistle gave him pleasure."

But the lesson it taught Franklin was cheap in the end. "As I grew up," he said, "and came into the world and observed the actions of men, I thought I met with many, very many, who gave too much for the whistle. In short, I conceive that a great part of the miseries of mankind are brought upon them by the false estimates they have made of the value of things, and by their giving too much for their whistles."

Gilbert and Sullivan paid too much for their whistle. So did Aunt Edith. So did Dale Carnegie-on many occasions. And so did the immortal Leo Tolstoy, author of two of the world's greatest novels, *War and Peace* and *Anna Karenina*. According to The Encyclopedia Britannica, Leo Tolstoy was, during the last twenty years of his life, "probably the most venerated man in the whole world." For twenty years before he died-from 1890 to

1910 an unending stream of admirers made pilgrimages to his home in order to catch a glimpse of his face, to hear the sound of his voice, or even touch the hem of his garment. Every sentence he uttered was taken down in a notebook, almost as if it were a "divine revelation". But when it came to living-to ordinary living-well, Tolstoy had even less sense at seventy than Franklin had at seven! He had no sense at all.

Here's what I mean. Tolstoy married a girl he loved very dearly. In fact, they were so happy together that they used to get on their knees and pray to God to let them continue their lives in such sheer, heavenly ecstasy. But the girl Tolstoy married was jealous by nature. She used to dress herself up as a peasant and spy on his movements, even out in the woods. They had fearful rows. She became so jealous, even of her own children, that she grabbed a gun and shot a hole in her daughter's photograph. She even rolled on the floor with an opium bottle held to her lips, and threatened to commit suicide, while the children huddled in a corner of the room and screamed with terror.

And what did Tolstoy do? Well, I don't blame the man for up and smashing the furniture-he had good provocation. But he did far worse than that. He kept a private diary! Yes, a diary, in which he placed all the blame on his wife! That was his "whistle"! He was determined to make sure that coming generations would exonerate him and put the blame on his wife. And what did his wife do, in answer to this? Why, she tore pages out of his diary and burned them, of course. She started a diary of her own, in which she made him the villain. She even wrote a novel, entitled *Whose Fault?* in which she depicted her husband as a household fiend and herself as a martyr.

All to what end? Why did these two people turn the only home they had into what Tolstoy himself called "a lunatic asylum"? Obviously, there were several reasons. One of those reasons was their burning desire to impress you and me. Yes, we are the posterity whose opinion they were worried about! Do we give a hoot in Hades about which one was to blame? No, we are too concerned with our own problems to waste a minute thinking about the Tolstoy's. What a price these two wretched people paid for their whistle! Fifty years of living in a veritable hell-just because neither of them had the sense to say: "Stop!" Because neither of them had enough judgment of values to say: "Let's put a stoploss order on this thing instantly. We are squandering our lives. Let's say 'Enough' now!"

Yes, I honestly believe that this is one of the greatest secrets to true peace of mind-a decent sense of values. And I believe we could annihilate fifty per cent of all our worries at once if we would develop a sort of private gold standard-a gold standard of what things are worth to us in terms of our lives.

So, to break the worry habit before it breaks you, here is Rule 5:

3.5

RULE 5

Whenever we are tempted to throw good money after bad in terms of human living, let's stop and ask ourselves these three Questions:

1. How much does this thing I am worrying about really matter to me?

2. At what point shall I set a "stop-loss" order on this worry - and forget it?

3. Exactly how much shall I pay for this whistle? Have I already paid more than it is worth?

✦ ✦ ✦

Chapter 11

Don't Try To Saw Sawdust

As I write this sentence, I can look out of my window and see some dinosaur tracks in my garden dinosaur tracks embedded in shale and stone. I purchased those dinosaur tracks from the Peabody Museum of Yale University; and I have a letter from the curator of the Peabody Museum, saying that those tracks were made 180 million years ago. Even a Mongolian idiot wouldn't dream of trying to go back 180 million years to change those tracks. Yet that would not be any more foolish than worrying because we can't go back and change what happened 180 seconds ago-and a lot of us are doing just that. To be sure, we may do something to modify the effects of what happened 180 seconds ago; but we can't possibly change the event that occurred then.

There is only one way on God's green footstool that the past can be constructive; and that is by calmly analysing our past mistakes and profiting by them-and forgetting them.

I know that is true; but have I always had the courage and sense to do it? To answer that question, let me tell you about a fantastic experience I had years ago. I let more than three hundred thousand dollars slip through my fingers without making a penny's profit. It happened like this: I launched a large-scale enterprise in adult education, opened branches in various cities, and spent money lavishly in overhead and advertising. I was so busy with teaching that I had neither the time nor the desire to look after finances. I was too naive to realise that I needed an astute business manager to watch expenses.

Finally, after about a year, I discovered a sobering and shocking truth. I discovered that in spite of our enormous intake, we had not netted any profit whatever. After discovering that, I should have done two things. First, I should have had the sense to do what George Washington Carver, the Negro scientist, did when he lost forty thousand dollars in a bank crash-the savings of a lifetime. When someone asked him if he knew he was bankrupt, he replied: "Yes, I heard"-and went on with his teaching. He wiped the loss out of his mind so completely that he never mentioned it again.

Here is the second thing I should have done: I should have analysed my mistakes and learned a lasting lesson.

But frankly, I didn't do either one of these things. Instead, I went into a tailspin of worry. For months I was in a daze. I lost sleep and I lost weight. Instead of learning a lesson from this enormous mistake, I went right ahead and did the same thing again on a smaller scale!

It is embarrassing for me to admit all this stupidity; but I discovered long ago that "it is easier to teach twenty what were good to be done than to be one of twenty to follow mine own teaching."

How I wish that I had had the privilege of attending the George Washington High School here in New York and studying under Mr. Brandwine-the same teacher who taught Allen Saunders, of 939 Woodycrest Avenue, Bronx, New York!

Mr. Saunders told me that the teacher of his hygiene class, Mr. Brandwine, taught him one of the most valuable lessons he had ever learned. "I was only in my teens," said Allen Saunders as he told me the story, "but I was a worrier even then. I used to stew and fret about the mistakes I had made. If I turned in an examination paper, I used to lie awake and chew my fingernails for fear I hadn't passed. I was always living over the things I had done, and wishing I'd done them differently; thinking over the things I had said, and wishing I'd said them better.

"Then one morning, our class filed into the science laboratory, and there was the teacher, Mr. Brandwine, with a bottle of milk prominently displayed on the edge of the desk. We all sat down, staring at the milk, and wondering what it had to do with the hygiene course he was teaching. Then, all of a sudden, Mr. Brandwine stood up, swept the bottle of milk with a crash into the sink-and shouted: 'Don't cry over spilt milk!'

"He then made us all come to the sink and look at the wreckage. 'Take a good look,' he told us, 'because I want you to remember this lesson the rest of your lives. That milk is gone you can see it's down the drain; and all the fussing and hair-pulling in the world won't bring back a drop of it. With a little thought and prevention, that milk might have been saved. But it's too late now-all we can do is write it off, forget it, and go on to the next thing.'

"That one little demonstration," Allen Saunders told me, "stuck with me long after I'd forgotten my solid geometry and Latin. In fact, it taught me more about practical living than anything else in my four years of high school. It taught me to keep from spilling milk if I could; but to forget it completely, once it was spilled and had gone down the drain."

Some readers are going to snort at the idea of making so much over a hackneyed proverb like "Don't cry over spilt milk." I know it is trite, commonplace, and a platitude. I know you have heard it a thousand times. But I also know that these hackneyed proverbs contain the very essence of the distilled wisdom of all ages. They have come out of the fiery experience of the human race and have been handed down through countless generations. If you were to read everything that has ever been written about worry by the great scholars of all time, you would never read anything more basic or more profound than such hackneyed proverbs as "Don't cross your bridges until you come to them" and "Don't cry over spilt milk." If we only applied those two proverbs-instead of snorting at them-we wouldn't need this book at all. In fact, if we applied most of the old proverbs, we would lead almost perfect lives. However, knowledge isn't power until it is applied; and the purpose of this book is not to tell you something new. The purpose of this book is to remind you of what you already know and to kick you in the shins and inspire you to do something about applying it.

I have always admired a man like the late Fred Fuller Shedd, who had a gift for stating an old truth in a new and picturesque way. He was editor of the Philadelphia Bulletin; and, while addressing a college graduating class, he asked: "How many of you have ever sawed wood? Let's see your hands." Most of them had. Then he inquired: "How many of you have ever sawed sawdust?" No hands went up.

"Of course, you can't saw sawdust!" Mr. Shedd exclaimed. "It's already sawed! And it's the same with the past. When you start worrying about things that are over and done with, you're merely trying to saw sawdust."

When Connie Mack, the grand old man of baseball, was eighty-one years old, I asked him if he had ever worried over games that were lost.

"Oh, yes, I used to," Connie Mack told me. "But I got over that foolishness long years ago. I found out it didn't get me anywhere at all. You can't grind any grain," he said, "with water that has already gone down the creek."

No, you can't grind any grain-and you can't saw any logs with water that has already gone down the creek. But you can saw wrinkles in your face and ulcers in your stomach.

I had dinner with Jack Dempsey last Thanksgiving; and he told me over the turkey and cranberry sauce about the fight in which he lost the heavyweight championship to Tunney Naturally, it was a blow to his ego. "In the midst of that fight," he told me, "I suddenly realised I had become an old man. ... At the end of the tenth round, I was still on my feet, but that was about all. My face was puffed and cut, and my eyes were nearly closed. ... I saw the referee raise Gene Tunney's hand in token of victory. ... I was no longer champion of the world. I started back in the rain-back through the crowd to my dressing-room. As I passed, some people tried to grab my hand. Others had tears in their eyes.

"A year later, I fought Tunney again. But it was no use. I was through forever. It was hard to keep from worrying about it all, but I said to myself: 'I'm not going to live in the past or cry over spilt milk. I am going to take this blow on the chin and not let it floor me.'"

And that is precisely what Jack Dempsey did. How? By saying to himself over and over: "I won't worry about the past"? No, that would merely have forced him to think of his past worries. He did it by accepting and writing off his defeat and then concentrating on plans for the future. He did it by running the Jack Dempsey Restaurant on Broadway and the Great Northern Hotel on 57th Street. He did it by promoting prize fights and giving boxing exhibitions. He did it by getting so busy on something constructive that he had neither the time nor the temptation to worry about the past. "I have had a better time during the last ten years," Jack Dempsey said, "than I had when I was champion."

As I read history and biography and observe people under trying circumstances, I am constantly astonished and inspired by some people's ability to write off their worries and tragedies and go on living fairly happy lives.

I once paid a visit to Sing Sing, and the thing that astonished me most was that the prisoners there appeared to be about as happy as the average person on the outside. I commented on it to Lewis E. Lawes-then warden of Sing Sing-and he told me that when criminals first arrive at Sing Sing, they are likely to be resentful and bitter. But after a few months, the majority of the more intelligent ones write off their misfortunes and settle down and accept prison life calmly and make the best of it. Warden Lawes told me about one Sing Sing prisoner- a gardener-who sang as he cultivated the vegetables and flowers inside the prison walls.

That Sing Sing prisoner who sang as he cultivated the flowers showed a lot more sense than most of us do. He knew that

The Moving Finger writes; and, having writ,
Moves on: nor all your Piety nor Wit
Shall lure it back to cancel half a Line,
Nor all your Tears wash out a Word of it.

So why waste the tears? Of course, we have been guilty of blunders and absurdities! And so what? Who hasn't? Even Napoleon lost one-third of all the important battles he fought. Perhaps our batting average is no worse than Napoleon's. Who knows?

And, anyhow, all the king's horses and all the king's men can't put the past together again. So let's remember Rule 6:

Part Three In A Nutshell - How To Break The Worry Habit Before It Breaks You

RULE 1 Crowd worry out of your mind by keeping busy. Plenty of action is one of the best therapies ever devised for curing "wibber gibbers".

RULE 2 Don't fuss about trifles. Don't permit little things-the mere termites of life-to ruin your happiness.

RULE 3 Use the law of averages to outlaw your worries. Ask yourself: "What are the odds against this thing's happening at all?"

RULE 4 Co-operate with the inevitable. If you know a circumstance is beyond your power to change or revise, say to yourself "It is so; it cannot be otherwise."

RULE 5 Put a "stop-loss" order on your worries. Decide just how much anxiety a thing may be worth and refuse to give it any more.

RULE 6 Let the past bury its dead. Don't saw sawdust.

✦ ✦ ✦

PART FOUR

Seven Ways To Cultivate A Mental Attitude That Will Bring You Peace And Happiness

Chapter 12

Eight Words That Can Transform Your Life

A Few years ago, I was asked to answer this question on a radio programme: "What is the biggest lesson you have ever learned?"

That was easy: by far the most vital lesson I have ever learned is the importance of what we think. If I knew what you think, I would know what you are. Our thoughts make us what we are. Our mental attitude is the X factor that determines our fate. Emerson said: "A man is what he thinks about all day long." ... How could he possibly be anything else?

I now know with a conviction beyond all doubt that the biggest problem you and I have to deal with-in fact, almost the only problem we have to deal with-is choosing the right thoughts. If we can do that, we will be on the highroad to solving all our problems. The great philosopher who ruled the Roman Empire, Marcus Aurelius, summed it up in eight words-eight words that can determine your destiny: "Our life is what our thoughts make it."

Yes, if we think happy thoughts, we will be happy. If we think miserable thoughts, we will be miserable. If we think fear thoughts, we will be fearful. If we think sickly thoughts, we will probably be ill. If we think failure, we will certainly fail. If we wallow in self-pity, everyone will want to shun us and avoid us. "You are not," said Norman Vincent Peale, "you are not what you think you are; but what you think, you are."

Am I advocating an habitual Pollyanna attitude toward all our problems? No, unfortunately, life isn't so simple as all that. But I am advocating that we assume a positive attitude instead of a negative attitude. In other words, we need to be concerned about our problems, but not worried. What is the difference between concern and worry? Let me illustrate. Every time I cross the traffic-jammed streets of New York, I am concerned about what I am doing-but not worried. Concern means realising what the problems are and calmly taking steps to meet them. Worrying means going around in maddening, futile circles.

A man can be concerned about his serious problems and still walk with his chin up and a carnation in his buttonhole. I have seen Lowell Thomas do just that. I once had the privilege of being associated with Lowell Thomas in presenting his famous films on the Allenby-Lawrence campaigns in World War I. He and his assistants had photographed the war on half a dozen fronts; and, best of all, had brought back a pictorial record of T. E. Lawrence and his colourful Arabian army, and a film record of Allenby's conquest of the Holy Land. His illustrated talks entitled "With Allenby in Palestine and Lawrence in Arabia" were a sensation in London-and around the world. The London opera season was postponed for six weeks so that he could continue telling his tale of high adventure and showing his pictures at Covent Garden Royal Opera House. After his sensational success in London came a triumphant tour of many countries. Then he spent two years preparing a film record of life in India and Afghanistan. After a lot of incredibly bad luck, the impossible happened: he found himself broke in London. I was with him at the time.

I remember we had to eat cheap meals at cheap restaurants. We couldn't have eaten even there if we had not borrowed money from a Scotsman-James McBey, the renowned artist. Here is the point of the story: even when Lowell Thomas was facing huge debts and severe disappointments, he was concerned, but not worried. He knew that if he let his reverses get him down, he would be worthless to everyone, including his creditors. So each morning before he started out, he bought a flower, put it in his buttonhole, and went swinging down Oxford Street with his head high and his step spirited. He thought positive, courageous thoughts and refused to let defeat defeat him.

To him, being licked was all part of the game-the useful training you had to expect if you wanted to get to the top.

Our mental attitude has an almost unbelievable effect even on our physical powers. The famous British psychiatrist, J. A. Hadfield, gives a striking illustration of that fact in his splendid book, *The Psychology of Power*. "I asked three men," he writes, "to submit themselves to test the effect of mental suggestion on their strength, which was measured by gripping a dynamometer." He told them to grip the dynamometer with all their might. He had them do this under three different sets of conditions.

When he tested them under normal waking conditions, their average grip was 101 pounds.

When he tested them after he had hypnotised them and told them that they were very weak, they could grip only 29 pounds -less than a third of their normal strength. (One of these men was a prize fighter; and when he was told under hypnosis that he was weak, he remarked that his arm felt "tiny, just like a baby's".)

When Captain Hadfield then tested these men a third time, telling them under hypnosis that they were very strong, they were able to grip an average of 142 pounds. When their minds were filled with positive thoughts of strength, they increased their actual physical powers almost five hundred per cent.

Such is the incredible power of our mental attitude.

To illustrate the magic power of thought, let me tell you one of the most astounding stories in the annals of America. I could write a book about it; but let's be brief. On a frosty October night, shortly after the close of the Civil War, a homeless, destitute woman, who was little more than a wanderer on the face of the earth, knocked at the door of "Mother" Webster, the wife of a retired sea captain, living in Amesbury, Massachusetts.

Opening the door, "Mother" Webster saw a frail little creature, "scarcely more than a hundred pounds of frightened skin and bones". The stranger, a Mrs. Glover, explained she was seeking a home where she could think and work out a great problem that absorbed her day and night.

"Why not stay here?" Mrs. Webster replied. "I'm all alone in this big house."

Mrs. Glover might have remained indefinitely with "Mother" Webster, if the latter's son-in-law, Bill Ellis, hadn't come up from New York for a vacation. When he discovered Mrs. Glover's presence, he shouted: "I'll have no vagabonds in this house"; and he shoved this homeless woman out of the door. A driving rain was falling. She stood shivering in the rain for a few minutes, and then started down the road, looking for shelter.

Here is the astonishing part of the story. That "vagabond" whom Bill Ellis put out of the house was destined to have as much influence on the thinking of the world as any other woman who ever walked this earth. She is now known to millions of devoted followers as Mary Baker Eddy-the founder of Christian Science.

Yet, until this time, she had known little in life except sickness, sorrow, and tragedy. Her first husband had died shortly after their marriage. Her second husband had deserted her and eloped with a married woman. He later died in a poor-house. She had only one child, a son; and she was forced, because of poverty, illness, and jealousy, to give him up when he was four years old. She lost all track of him and never saw him again for thirty-one years.

Because of her own ill health, Mrs. Eddy had been interested for years in what she called "the science of mind healing". But the dramatic turning point in her life occurred in Lynn, Massachusetts. Walking downtown one cold day, she slipped and fell on the icy pavement-and was knocked unconscious. Her spine was so injured that she was convulsed with spasms. Even the doctor expected her to die. If by some miracle she lived, he declared that she would never walk again.

Lying on what was supposed to be her deathbed, Mary Baker Eddy opened her Bible, and was led, she declared, by divine guidance to read these words from Saint Matthew: "And, behold, they brought to him a man sick of the palsy, lying on a bed: and Jesus ... said unto the sick of the palsy: Son, be of good cheer;

thy sins be forgiven thee. ... Arise, take up thy bed, and go unto thine house. And he arose, and departed to his house."

These words of Jesus, she declared, produced within her such a strength, such a faith, such a surge of healing power, that she "immediately got out of bed and walked".

"That experience," Mrs. Eddy declared, "was the falling apple that led me to the discovery of how to be well myself, and how to make others so. ... I gained the scientific certainty that all causation was Mind, and every effect a mental phenomenon."

Such was the way in which Mary Baker Eddy became the founder and high priestess of a new religion: Christian Science -the only great religious faith ever established by a woman- a religion that has encircled the globe.

You are probably saying to yourself by now: "This man Carnegie is proselytising for Christian Science." No. You are wrong. I am not a Christian Scientist. But the longer I live, the more deeply I am convinced of the tremendous power of thought. As a result of thirty-five years spent in teaching adults, I know men and women can banish worry, fear, and various kind of illness, and can transform their lives by changing their thoughts. I know! I know! ! I know! ! ! I have seen such incredible transformations performed hundreds of times. I have seen them so often that I no longer wonder at them.

For example, one of these transformations happened to one of my students, Frank J. Whaley, of 1469 West Idaho Street, Saint Paul, Minnesota. He had a nervous breakdown. What brought it on? Worry. Frank Whaley tells me: "I worried about everything: I worried because I was too thin; because I thought I was losing my hair; because I feared I would never make enough money to get married; because I felt I would never make a good father; because I feared I was losing the girl I wanted to marry; because I felt I was not living a good life. I worried about the impression I was making on other people. I worried because I thought I had stomach ulcers. I could no longer work; I gave up my job. I built up tension inside me until I was like a boiler without a safety valve. The pressure got so unbearable that something had to give-and it did. If you have never had a nervous breakdown, pray God that you never do, for no pain of the body can exceed the excruciating pain of an agonised mind.

"My breakdown was so severe that I couldn't talk even to my own family. I had no control over my thoughts. I was filled with fear. I would jump at the slightest noise. I avoided everybody. I would break out crying for no apparent reason at all.

"Every day was one of agony. I felt that I was deserted by everybody-even God. I was tempted to jump into the river and end it all.

"I decided instead to take a trip to Florida, hoping that a change of scene would help me. As I stepped on the train, my father handed me a letter and told me not to open it until I reached Florida. I landed in Florida during the height of the tourist season. Since I couldn't get in a hotel, I rented a sleeping room in a garage. I tried to get a job on a tramp freighter out of Miami, but had no luck. So I spent my time at the beach. I was more wretched in Florida than I had been at home; so I opened the envelope to see what Dad had written. His note said: 'Son, you are 1,500 miles from home, and you don't feel any different, do you? I knew you wouldn't, because you took with you the one thing that is the cause of all your trouble, that is, yourself. There is nothing wrong with either your body or your mind. It is not the situations you have met that have thrown you; it is what you think of these situations. "As a man thinketh in his heart, so is he." When you realise that, son, come home, for you will be cured.'

"Dad's letter made me angry. I was looking for sympathy, not instruction. I was so mad that I decided then and there that I would never go home. That night as I was walking down one of the side streets of Miami, I came to a church where services were going on. Having no place to go, I drifted in and listened to a sermon on the text: 'He who conquers his spirit is mightier than he who taketh a city.' Sitting in the sanctity of the house of God and hearing the same thoughts that my Dad had written in his letter-all this swept the accumulated litter out of my brain. I was able to think clearly and sensibly for the first time in my life. I realised what a fool I had been. I was shocked to see myself in my true light: here I was, wanting to change the whole world and everyone in it- when the only thing that needed changing was the focus of the lens of the camera which was my mind.

"The next morning I packed and started home. A week later I was back on the job. Four months later I married the girl I had been afraid of losing. We now have a happy family of five children. God has been good to me both materially and mentally. At the time of the breakdown I was a night foreman of a small department handling eighteen people. I am now superintendent of carton manufacture in charge of over four hundred and fifty people. Life is much fuller and friendlier. I believe I appreciate the true values of life now. When moments of uneasiness try to creep in (as they will in everyone's life) I tell myself to get that camera back in focus, and everything is O.K.

"I can honestly say that I am glad I had the breakdown, because I found out the hard way what power our thoughts can have over our mind and our body. Now I can make my thoughts work for me instead of against me. I can see now that Dad was right when he said it wasn't outward situations that had caused all my suffering, but what I thought of those situations. And as soon as I realised that, I was cured-and stayed cured." Such was the experience of Frank J. Whaley.

I am deeply convinced that our peace of mind and the joy we get out of living depends not on where we are, or what we have, or who we are, but solely upon our mental attitude. Outward conditions have very little to do with it. For example, let's take the case of old John Brown, who was hanged for seizing the United States arsenal at Harpers Ferry and trying to incite the slaves to rebellion. He rode away to the gallows, sitting on his coffin. The jailer who rode beside him was nervous and worried. But old John Brown was calm and cool. Looking up at the Blue Ridge mountains of Virginia, he exclaimed: "What a beautiful country! I never had an opportunity to really see it before."

Or take the case of Robert Falcon Scott and his companions-the first Englishman ever to reach the South Pole. Their return trip was probably the cruelest journey ever undertaken by man. Their food was gone-and so was their fuel. They could no longer march because a howling blizzard roared down over the rim of the earth for eleven days and nights-a wind so fierce and sharp that it cut ridges in the polar ice. Scott and his companions knew they were going to die; and they had brought a quantity of opium along for just such an emergency. A big dose of opium, and they could all lie down to pleasant dreams, never to wake again. But they ignored the drug, and died "singing ringing songs of cheer". We know they did because of a farewell letter found with their frozen bodies by a searching party, eight months later.

Yes, if we cherish creative thoughts of courage and calmness, we can enjoy the scenery while sitting on our coffin, riding to the gallows; or we can fill our tents with "ringing songs of cheer", while starving and freezing to death.

Milton in his blindness discovered that same truth three hundred years ago:

The mind is its own place, and in itself
Can make a heaven of Hell, a hell of Heaven.

Napoleon and Helen Keller are perfect illustrations of Milton's statement: Napoleon had everything men usually crave-glory, power, riches-yet he said at St. Helena: "I have never known six happy days in my life"; while Helen Keller- blind, deaf, dumb-declared: "I have found life so beautiful."

If half a century of living has taught me anything at all, it has taught me that "Nothing can bring you peace but yourself."

I am merely trying to repeat what Emerson said so well in the closing words of his essay on "Self- Reliance" : "A political victory, a rise in rents, the recovery of your sick, or the return of your absent friend, or some other quite external event, raises your spirits, and you think good days are preparing for you. Do not believe it. It can never be so. Nothing can bring you peace but yourself."

Epictetus, the great Stoic philosopher, warned that we ought to be more concerned about removing wrong thoughts from the mind than about removing "tumours and abscesses from the body."

Epictetus said that nineteen centuries ago, but modern medicine would back him up. Dr. G. Canby Robinson declared that four out of five patients admitted to Johns Hopkins Hospital were suffering from conditions brought on in part by emotional strains and stresses. This was often true even in cases of organic disturbances. "Eventually," he declared, "these trace back to maladjustments to life and its problems."

Montaigne, the great French philosopher, adopted these seventeen words as the motto of his life: "A man is not hurt so much by what happens, as by his opinion of what happens." And our opinion of what happens is entirely up to us.

What do I mean? Have I the colossal effrontery to tell you to your face-when you are mowed down by troubles, and your nerves are sticking out like wires and curling up at the ends-have I the colossal effrontery to tell you that, under those conditions, you can change your mental attitude by an effort of will? Yes, I mean precisely that! And that is not all. I am going to show you how to do it. It may take a little effort, but the secret is simple.

William James, who has never been topped in his knowledge of practical psychology, once made this observation: "Action seems to follow feeling, but really action and feeling go together; and by regulating the action, which is under the more direct control of the will, we can indirectly regulate the feeling, which is not."

In other words, William James tells us that we cannot instantly change our emotions just by "making up our minds to"-but that we can change our actions. And that when we change our actions, we will automatically change our feelings.

"Thus," he explains, "The sovereign voluntary path to cheerfulness, if your cheerfulness be lost, is to sit up cheerfully and to act and speak as if cheerfulness were already there."

Does that simple trick work? It works like plastic surgery! Try it yourself. Put a big, broad, honest-to- God smile on your face; throw back your shoulders; take a good, deep breath; and sing a snatch of song. If you can't sing, whistle. If you can't whistle, hum. You will quickly discover what William James was talking about-that it is physically impossible to remain blue or depressed while you are acting out the symptoms of being radiantly happy!

This is one of the little basic truths of nature that can easily work miracles in all our lives. I know a woman in California -I won't mention her name-who could wipe out all of her miseries in twenty four hours if only she knew this secret. She's old, and she's a widow-that's sad, I admit-but does she try to act happy? No; if you ask her how she is feeling, she says: "Oh, I'm all right"-but the expression on her face and the whine in her voice say: "Oh, God, if you only knew the troubles I've seen!" She seems to reproach you for being happy in her presence. Hundreds of women are worse off that she is: her husband left her enough insurance to last the rest of her life, and she has married children to give her a home. But I've rarely seen her smile.

She complains that all three of her sons-in-law are stingy and selfish-although she is a guest in their homes for months at a time. And she complains that her daughters never give her presents-although she hoards her own money carefully, "for my old age". She is a blight on herself and her unfortunate family! But does it have to be so? That is the pity of it-she could change herself from a miserable, bitter, and unhappy old woman into an honoured and beloved member of the family-if she wanted to change. And all she would have to do to work this transformation would be to start acting cheerful; start acting as though she had a little love to give away-instead of squandering it all on her own unhappy and embittered self.

I know a man in Indiana-H. J. Englert, of 1335th Street, Tell City, Indiana-who is still alive today because he discovered this secret. Ten years ago Mr. Englert had a case of scarlet fever; and when he recovered, he found he had developed nephritis, a kidney disease. He tried all kinds of doctors, "even quacks", he informs me, but nothing could cure him.

Then, a short time ago, he got other complications. His blood pressure soared. He went to a doctor, and was told that his blood pressure was hitting the top at 214. He was told that it was fatal-that the condition was progressive, and he had better put his affairs in order at once.

"I went home," he says, "and made sure that my insurance was all paid up, and then I apologised to my Maker for all my mistakes, and settled down to gloomy meditations.

"I made everyone unhappy. My wife and family were miserable, and I was buried deep in depression myself. However, after a week of wallowing in self-pity, I said to myself: 'You're acting like a fool! You may not die for a year yet, so why not try to be happy while you're here?'

"I threw back my shoulders, put a smile on my face, and attempted to act as though everything were normal. I admit it was an effort at first-but I forced myself to be pleasant and cheerful; and this not only helped my family, but it also helped me.

"The first thing I knew, I began to feel better-almost as well as I pretended to feel! The improvement went on. And today-months after I was supposed to be in my grave-I am not only happy, well, and alive, but my blood pressure is down! I know one thing for certain: the doctor's prediction would certainly have come true if I had gone on thinking 'dying' thoughts of defeat. But I gave my body a chance to heal itself, by nothing in the world but a change of mental attitude!"

Let me ask you a question: If merely acting cheerful and thinking positive thoughts of health and courage can save this man's life, why should you and I tolerate for one minute more our minor glooms and depressions? Why make ourselves, and everyone around us, unhappy and blue, when it is possible for us to start creating happiness by merely acting cheerful?

Years ago, I read a little book that had a lasting and profound effect on my life. It was called *As a Man Thinketh* (*) by James Lane Allen, and here's what it said:

---

* Fowler & Co. Ltd.

"A man will find that as he alters his thoughts towards things and other people, things and other people will alter towards him. ... Let a man radically alter his thoughts, and he will be astonished at the rapid transformation it will effect in the material conditions of his life. Men do not attract that which they want, but that which they are. ... The divinity that shapes our ends is in ourselves. It is our very self. ... All that a man achieves is the direct result of his own thoughts. ... A man can only rise, conquer and achieve by lifting up his thoughts. He can only remain weak and abject and miserable by refusing to lift up his thoughts."

According to the book of Genesis, the Creator gave man dominion over the whole wide earth. A mighty big present. But I am not interested in any such super-royal prerogatives. All I desire is dominion over myself-dominion over my thoughts; dominion over my fears; dominion over my mind and over my spirit. And the wonderful thing is that I know that I can attain this dominion to an astonishing degree, any time I want to, by merely controlling my actions-which in turn control my reactions.

So let us remember these words of William James: "Much of what we call Evil ... can often be converted into a bracing and tonic good by a simple change of the sufferer's inner attitude from one of fear to one of fight."

Let's fight for our happiness!

Let's fight for our happiness by following a daily programme of cheerful and constructive thinking. Here is such a programme. It is entitled "Just for Today". I found this programme so inspiring that I gave away hundreds of copies. It was written thirty-six years ago by the late Sibyl F. Partridge. If you and I follow it, we will eliminate most of our worries and increase immeasurably our portion of what the French call la joie de vivre.

Just For Today

1. Just for today I will be happy. This assumes that what Abraham Lincoln said is true, that "most folks are about as happy as they make up their minds to be." Happiness is from within; it is not a matter of externals.

2 Just for today I will try to adjust myself to what is, and not try to adjust everything to my own desires. I will take my family, my business, and my luck as they come and fit myself to them.

3 Just for today I will take care of my body. I will exercise it, care for it, nourish it, not abuse it nor neglect it, so that it will be a perfect machine for my bidding.

4 Just for today I will try to strengthen my mind. I will learn something useful. I will not be a mental loafer. I will read something that requires effort, thought and concentration.

5 Just for today I will exercise my soul in three ways: I will do somebody a good turn and not get found out. I will do at least two things I don't want to do, as William James suggests, just for exercise.

6 Just for today I will be agreeable. I will look as well as I can, dress as becomingly as possible, talk low, act courteously, be liberal with praise, criticise not at all, nor find fault with anything and not try to regulate nor improve anyone.

7 Just for today I will try to live through this day only, not to tackle my whole life problem at once. I can do things for twelve hours that would appall me if I had to keep them up for a lifetime.

8 Just for today I will have a programme. I will write down what I expect to do every hour. I may not follow it exactly, but I will have it. It will eliminate two pests, hurry and indecision.

9 Just for today I will have a quiet half-hour all by myself and relax. In this half-hour sometimes I will think of God, so as to get a little more perspective into my life.

10 Just for today I will be unafraid, especially I will not be afraid to be happy, to enjoy what is beautiful, to love, and to believe that those I love, love me.

If we want to develop a mental attitude that will bring us peace and happiness, here is Rule 1:

4.1 RULE 1

Think and act cheerfully, and you will feel cheerful.

⁂

Chapter 13

The High Cost Of Getting Even

One night, years ago, as I was travelling through Yellowstone Park, I sat with other tourists on bleachers facing a dense growth of pine and spruce. Presently the animal which we had been waiting to see, the terror of the forests, the grizzly bear, strode out into the glare of the lights and began devouring the garbage that had been dumped there from the kitchen of one of the park hotels. A forest ranger, Major Martindale, sat on a horse and talked to the excited tourists about bears. He told us that the grizzly bear can whip any other animal in the Western world, with the possible exception of the buffalo and the Kadiak bear; yet I noticed that night that there was one animal, and only one, that the grizzly permitted to come out of the forest and eat with him under the glare of the lights: a skunk. The grizzly knew that he could liquidate a skunk with one swipe of his mighty paw. Why didn't he do it? Because he had found from experience that it didn't pay.

I found that out, too. As a farm boy, I trapped four-legged skunks along the hedgerows in Missouri; and, as a man, I encountered a few two-legged skunks on the sidewalks of New York. I have found from sad experience that it doesn't pay to stir up either variety.

When we hate our enemies, we are giving them power over us: power over our sleep, our appetites, our blood pressure, our health, and our happiness. Our enemies would dance with joy if only they knew how they were worrying us, lacerating us and getting even with us! Our hate is not hurting them, but our hate is turning our own days and nights into a hellish turmoil.

Who do you suppose said this: "If selfish people try to take advantage of you, cross them off your list, but don't try to get even. When you try to get even, you hurt yourself more than you hurt the other fellow"? ... Those words sound as if they might have been uttered by some starry-eyed idealist. But they weren't. Those words appeared in a bulletin issued by the Police Department of Milwaukee.

How will trying to get even hurt you? In many ways. According to Life magazine, it may even wreck your health. "The chief personality characteristic of persons with hypertension [high blood pressure] is resentment," said Life. "When resentment is chronic, chronic hypertension and heart trouble follow."

So you see that when Jesus said: "Love your enemies", He was not only preaching sound ethics. He was also preaching twentieth-century medicine. When He said: "Forgive seventy time seven", Jesus was telling you and me how to keep from having high blood pressure, heart trouble, stomach ulcers, and many other ailments.

A friend of mine recently had a serious heart attack. Her physician put her to bed and ordered her to refuse to get angry about anything, no matter what happened. Physicians know that if you have a weak heart, a fit of anger can kill you. Did I say can kill you? A fit of anger did kill a restaurant owner in Spokane, Washington, a few years ago. I have in front of me now a letter from Jerry Swartout, chief of the Police Department, Spokane, Washington, saying: "A few years ago, William Falkaber, a man of sixty-eight who owned a cafe here in Spokane, killed himself by flying into a rage because his cook insisted on drinking coffee out of his saucer. The cafe owner was so indignant that he grabbed a revolver and started to chase the cook and fell dead from heart failure-with his hand still gripping the gun. The coroner's report declared that anger had caused the heart failure."

When Jesus said: "Love your enemies", He was also telling us how to improve our looks. I know women-and so do you-whose faces have been wrinkled and hardened by hate and disfigured by resentment. All the beauty treatments in Christendom won't improve their looks half so much as would a heart full of forgiveness, tenderness, and love.

Hatred destroys our ability to enjoy even our food. The Bible puts it this way "Better is a dinner of herbs where love is, than a stalled ox and hatred therewith."

Wouldn't our enemies rub their hands with glee if they knew that our hate for them was exhausting us, making us tired and nervous, ruining our looks, giving us heart trouble, and probably shortening our lives?

Even if we can't love our enemies, let's at least love ourselves. Let's love ourselves so much that we won't permit our enemies to control our happiness, our health and our looks. As Shakespeare put it:

Heat not a furnace for your foe so hot
That it do singe yourself.

When Jesus said that we should forgive our enemies "seventy times seven", He was also preaching sound business. For example, I have before me as I write a letter I received from George Rona, Fradegata'n 24, Uppsala, Sweden. For years, George Rona was an attorney in Vienna; but during the Second World War, he fled to Sweden. He had no money, needed work badly. Since he could speak and write several languages, he hoped to get a position as correspondent for some firm engaged in importing or exporting. Most of the firms replied that they had no need of such services because of the war, but they would keep his name on file ... and so on. One man, however, wrote George Rona a letter saying: "What you imagine about my business is not true. You are both wrong and foolish. I do not need any correspondent. Even if I did need one, I wouldn't hire you because you can't even write good Swedish. Your letter is full of mistakes."

When George Rona read that letter, he was as mad as Donald Duck. What did this Swede mean by telling him he couldn't write the language! Why, the letter that this Swede himself had written was full of mistakes! So George Rona wrote a letter that was calculated to burn this man up. Then he paused. He said to himself: "Wait a minute, now. How do I know this man isn't right? I have studied Swedish, but it's not my native language, so maybe I do make mistakes I don't know anything about. If I do, then I certainly have to study harder if I ever hope to get a job. This man has possibly done me a favour, even though he didn't mean to. The mere fact that he expressed himself in disagreeable terms doesn't alter my debt to him. Therefore, I am going to write him and thank him for what he has done."

So George Rona tore up the scorching letter he had already written, and wrote another that said: "It was kind of you to go to the trouble of writing to me, especially when you do not need a correspondent. I am sorry I was mistaken about your firm. The reason that I wrote you was that I made inquiry and your name was given me as a leader in your field. I did not know I had made grammatical errors in my letter. I am sorry and ashamed of myself. I will now apply myself more diligently to the study of the Swedish language and try to correct my mistakes. I want to thank you for helping me get started on the road to self-improvement."

Within a few days, George Rona got a letter from this man, asking Rona to come to see him. Rona went-and got a job. George Rona discovered for himself that "a soft answer turneth away wrath".

We may not be saintly enough to love our enemies, but, for the sake of our own health and happiness, let's at least forgive them and forget them. That is the smart thing to do. "To be wronged or robbed," said Confucius, "is nothing unless you continue to remember it." I once asked General Eisenhower's son, John, if his father ever nourished resentments. "No," he replied, "Dad never wastes a minute thinking about people he doesn't like."

There is an old saying that a man is a fool who can't be angry, but a man is wise who won't be angry.

That was the policy of William J. Gaynor, former Mayor of New York. Bitterly denounced by the yellow press, he was shot by a maniac and almost killed. As he lay in the hospital, fighting for his life, he said: "Every night, I forgive everything and everybody." Is that too idealistic? Too much sweetness and light? If so, let's turn for counsel to the great German philosopher, Schopenhauer, author of *Studies in Pessimism*.

He regarded life as a futile and painful adventure. Gloom dripped from him as he walked; yet out of the depths of his despair, Schopenhauer cried: "If possible, no animosity should be felt for anyone."

I once asked Bernard Baruch-the man who was the trusted adviser to six Presidents: Wilson, Harding, Coolidge, Hoover, Roosevelt, and Truman-whether he was ever disturbed by the attacks of his enemies. "No man can humiliate me or disturb me," he replied. "I won't let him."

No one can humiliate or disturb you and me, either-unless we let him.

Sticks and stones may break my bones,
But words can never hurt me.

"Throughout the ages mankind has burned its candles before those Christlike individuals who bore no malice against their enemies. I have often stood in the Jasper National Park, in Canada, and gazed upon one of the most beautiful mountains in the Western world-a mountain named in honour of Edith Cavell, the British nurse who went to her death like a saint before a German firing squad on October 12, 1915. Her crime? She had hidden and fed and nursed wounded French and English soldiers in her Belgian home, and had helped them escape into Holland. As the English chaplain entered her cell in the military prison in Brussels that October morning, to prepare her for death, Edith Cavell uttered two sentences that have been preserved in bronze and granite: "I realise that patriotism is not enough. I must have no hatred or bitterness toward anyone." Four years later, her body was removed to England and memorial services were held in Westminster Abbey. Today, a granite statue stands opposite the National Portrait Gallery in London-a statue of one of England's immortals. "I realise that patriotism is not enough. I must have no hatred or bitterness toward anyone."

One sure way to forgive and forget our enemies is to become absorbed in some cause infinitely bigger than ourselves. Then the insults and the enmities we encounter won't matter because we will be oblivious of everything but our cause. As an example, let's take an intensely dramatic event that was about to take place in the pine woods of Mississippi back in 1918. A lynching! Laurence Jones, a coloured teacher and preacher, was about to be lynched. A few years ago, I visited the school that Laurence Jones founded-the Piney Woods Country School-and I spoke before the student body. That school is nationally known today, but the incident I am going to relate occurred long before that. It occurred back in the highly emotional days of the First World War. A rumour had spread through central Mississippi that the Germans were arousing the Negroes and inciting them to rebellion. Laurence Jones, the man who was about to be lynched, was, as I have already said, a Negro himself and was accused of helping to arouse his race to insurrection. A group of white men-pausing outside the church-had heard Laurence Jones shouting to his congregation: "Life is a battle in which every Negro must gird on his armour and fight to survive and succeed."

"Fight!" "Armour!" Enough! Galloping off into the night, these excited young men recruited a mob, returned to the church, put a rope round the preacher, dragged him for a mile up the road, stood him on a heap of faggots, lighted matches, and were ready to hang him and burn him at the same time, when someone shouted: "Let's make the blankety-blank-blank talk before he burns. Speech! Speech!" Laurence Jones, standing on the faggots, spoke with a rope around his neck, spoke for his life and his cause. He had been graduated from the University of Iowa in 1907. His sterling character, his scholarship and his musical ability had made him popular with both the students and the faculty. Upon graduation, he had turned down the offer of a hotel man to set him up in business, and had turned down the offer of a wealthy man to finance his musical education. Why? Because he was on fire with a vision. Reading the story of Booker T. Washington's life, he had been inspired to devote his own life to educating the poverty-stricken, illiterate members of his race. So he went to the most backward belt he could find in the South a spot twenty-five miles south of Jackson, Mississippi. Pawning his watch for $1.65, he started his school in the open woods with a stump for a desk. Laurence Jones told these angry men who were waiting to lynch him of the struggle he had had to educate these unschooled boys and girls and to train them to be good farmers, mechanics, cooks, housekeepers. He told of the white men who had helped him in his struggle to establish Piney Woods Country School-white men who had given him land, lumber, and pigs, cows and money, to help him carry on his educational work.

When Laurence Jones was asked afterward if he didn't hate the men who had dragged him up the road to hang him and burn him, he replied that he was too busy with his cause to hate-too absorbed in something bigger than himself. "I have no time to quarrel," he said, "no time for regrets, and no man can force me to stoop low enough to hate him."

As Laurence Jones talked with sincere and moving eloquence as he pleaded, not for himself but his cause, the mob began to soften. Finally, an old Confederate veteran in the crowd said: "I believe this boy is telling the truth. I know the white men whose names he has mentioned. He is doing a fine work. We have made a mistake. We ought to help him instead of hang him." The Confederate veteran passed his hat through the crowd and raised a gift of fifty-two dollars and forty cents from the very men who had gathered there to hang the founder of Piney Woods Country School-the man who said: "I have no time to quarrel, no time for regrets, and no man can force me to stoop low enough to hate him."

Epictetus pointed out nineteen centuries ago that we reap what we sow and that somehow fate almost always makes us pay for our malefactions. "In the long run," said Epictetus, "every man will pay the penalty for his own misdeeds. The man who remembers this will be angry with no one, indignant with no one, revile no one, blame no one, offend no one, hate no one."

Probably no other man in American history was ever more denounced and hated and double-crossed than Lincoln. Yet Lincoln, according to Herndon's classic biography, "never judged men by his like or dislike for them. If any given act was to be performed, he could understand that his enemy could do it just as well as anyone. If a man had maligned him or been guilty of personal ill-treatment, and was the fittest man for the place, Lincoln would give him that place, just as soon as he would give it to a friend. ... I do not think he ever removed a man because he was his enemy or because he disliked him."

Lincoln was denounced and insulted by some of the very men he had appointed to positions of high power-men like McClellan, Seward, Stanton, and Chase. Yet Lincoln believed, according to Herndon, his law partner, that "No man was to be eulogised for what he did; or censured for what he did or did not do," because "all of us are the children of conditions, of circumstances, of environment, of education, of acquired habits and of heredity moulding men as they are and will forever be."

Perhaps Lincoln was right. If you and I had inherited the same physical, mental, and emotional characteristics that our enemies have inherited, and if life had done to us what it has done to them, we would act exactly as they do. We couldn't possibly do anything else. As Clarence Darrow used to say: "To know all is to understand all, and this leaves no room for judgment and condemnation." So instead of hating our enemies, let's pity them and thank God that life has not made us what they are. Instead of heaping condemnation and revenge upon our enemies, let's give them our understanding, our sympathy, our help, our forgiveness, and our prayers."

I was brought up in a family which read the Scriptures or repeated a verse from the Bible each night and then knelt down and said "family prayers". I can still hear my father, in a lonely Missouri farmhouse, repeating those words of Jesus- words that will continue to be repeated as long as man cherishes his ideals: "Love your enemies, bless them that curse you, do good to them that hate you, and pray for them which despitefully use you, and persecute you."

My father tried to live those words of Jesus; and they gave him an inner peace that the captains and the kings of earth have often sought for in vain.

To cultivate a mental attitude that will bring you peace and happiness, remember that Rule 2 is:

4.2

RULE 2

Let's never try to get even with our enemies, because if we do we will hurt ourselves far more than we hurt them.

Let's do as General Eisenhower does: let's never waste a minute thinking about people we don't like.

✣ ✣ ✣

Chapter 14

If You Do This, You Will Never Worry About Ingratitude

I recently met a businessman in Texas who was burned up with indignation. I was warned that he would tell me about it within fifteen minutes after I met him. He did. The incident he was angry about had occurred eleven months previously, but he was still burned up about it. He couldn't speak of anything else. He had given his thirty-four employees ten thousand dollars in Christmas bonuses, approximately three hundred dollars each-and no one had thanked him. "I am sorry," he complained bitterly, "that I ever gave them a penny!"

"An angry man," said Confucius, "is always full of poison." This man was so full of poison that I honestly pitied him. He was about sixty years old. Now, life-insurance companies figure that, on the average, we will live slightly more than two-thirds of the difference between our present age and eighty. So this man-if he was lucky-probably had about fourteen or fifteen years to live. Yet he had already wasted almost one of his few remaining years by his bitterness and resentment over an event that was past and gone. I pitied him.

Instead of wallowing in resentment and self-pity, he might have asked himself why he didn't get any appreciation. Maybe he had underpaid and overworked his employees. Maybe they considered a Christmas bonus not a gift, but something they had earned. Maybe he was so critical and unapproachable that no one dared or cared to thank him. Maybe they felt he gave the bonus because most of the profits were going for taxes, anyway.

On the other hand, maybe the employees were selfish, mean, and ill-mannered. Maybe this. Maybe that. I don't know any more about it than you do. But I do know what Dr. Samuel

Johnson said: "Gratitude is a fruit of great cultivation. You do not find it among gross people."

Here is the point I am trying to make: this man made the human and distressing mistake of expecting gratitude. He just didn't know human nature.

If you saved a man's life, would you expect him to be grateful? You might-but Samuel Leibowitz, who was a famous criminal lawyer before he became a judge, saved seventy-eight men from going to the electric chair! How many of these men, do you suppose, stopped to thank Samuel Leibowitz, or ever took the trouble to send him a Christmas card? How many? Guess. ... That's right-none.

Christ healed ten lepers in one afternoon-but how many of those lepers even stopped to thank Him? Only one. Look it up in Saint Luke. When Christ turned around to His disciples and asked: "Where are the other nine?" they had all run away. Disappeared without thanks! Let me ask you a question: Why should you and I-or this businessman in Texas-expect more thanks for our small favours than was given Jesus Christ?

And when it comes to money matters! Well, that is even more hopeless. Charles Schwab told me that he had once saved a bank cashier who had speculated in the stock market with funds belonging to the bank. Schwab put up the money to save this man from going to the penitentiary. Was the cashier grateful? Oh, yes, for a little while. Then he turned against Schwab and reviled him and denounced him-the very man who had kept him out of jail!

If you gave one of your relatives a million dollars, would you expect him to be grateful? Andrew Carnegie did just that. But if Andrew Carnegie had come back from the grave a little while later, he would have been shocked to find this relative cursing him! Why? Because Old Andy had left 365 million dollars to public charities-and had "cut him off with one measly million," as he put it.

That's how it goes. Human nature has always been human nature-and it probably won't change in your lifetime. So why not accept it? Why not be as realistic about it as was old Marcus Aurelius, one of the wisest men who ever ruled the Roman Empire. He wrote in his diary one day: "I am going to meet people today who talk too much-people who are selfish, egotistical, ungrateful. But I won't be surprised or disturbed, for I couldn't imagine a world without such people." That makes sense, doesn't it? If you and I go around grumbling about ingratitude, who is to blame? Is it human nature-or is it our ignorance of human nature? Let's not expect gratitude. Then, if we get some occasionally, it will come as a delightful surprise. If we don't get it, we won't be disturbed.

Here is the first point I am trying to make in this chapter: It is natural for people to forget to be grateful; so, if we go around expecting gratitude, we are headed straight for a lot of heartaches.

I know a woman in New York who is always complaining because she is lonely. Not one of her relatives wants to go near her-and no wonder. If you visit her, she will tell you for hours what she did for her nieces when they were children: she nursed them through the measles and the mumps and the whooping-cough; she boarded them for years; she helped to send one of them through business school, and she made a home for the other until she got married.

Do the nieces come to see her? Oh, yes, now and then, out of a spirit of duty. But they dread these visits. They know they will have to sit and listen for hours to half-veiled reproaches. They will be treated to an endless litany of bitter complaints and self-pitying sighs. And when this woman can no longer bludgeon, browbeat, or bully her nieces into coming to see her, she has one of her "spells". She develops a heart attack.

Is the heart attack real? Oh, yes. The doctors say she has "a nervous heart", suffers from palpitations. But the doctors also say they can do nothing for her-her trouble is emotional.

What this woman really wants is love and attention. But she calls it "gratitude". And she will never get gratitude or love, because she demands it. She thinks it's her due.

There are thousands of women like her, women who are ill from "ingratitude", loneliness, and neglect. They long to be loved; but the only way in this world that they can ever hope to be loved is to stop asking for it and to start pouring out love without hope of return.

Does that sound like sheer, impractical, visionary idealism? It isn't. It is just horse sense. It is a good way for you and me to find the happiness we long for. I know. I have seen it happen right in my own family. My own mother and father gave for the joy of helping others. We were poor-always overwhelmed by debts. Yet, poor as we were, my father and mother always managed to send money every year to an orphans' home-the Christian Home in Council Bluffs, Iowa. Mother and Father never visited that home. Probably no one thanked them for their gifts-except by letter-but they were richly repaid, for they had the joy of helping little children-without wishing for or expecting any gratitude in return.

After I left home, I would always send Father and Mother a cheque at Christmas and urge them to indulge in a few luxuries for themselves. But they rarely did. When I came home a few days before Christmas, Father would tell me of the coal and groceries they had bought for some "widder woman" in town who had a lot of children and no money to buy food and fuel. What joy they got out of these gifts-the joy of giving without accepting anything whatever in return!

I believe my father would almost have qualified for Aristotle's description of the ideal man-the man most worthy of being happy. "The ideal man," said Aristotle, "takes joy in doing favours for others; but he feels ashamed to have others do favours for him. For it is a mark of superiority to confer a kindness; but it is a mark of inferiority to receive it."

Here is the second point I am trying to make in this chapter: If we want to find happiness, let's stop thinking about gratitude or ingratitude and give for the inner joy of giving.

Parents have been tearing their hair about the ingratitude of children for ten thousand years. Even Shakespeare's King Lear cried out: "How sharper than a serpent's tooth it is to have a thankless child!" But why should children be thankful-unless we train them to be? Ingratitude is natural-like weeds. Gratitude is like a rose. It has to be fed and watered and cultivated and loved and protected.

If our children are ungrateful, who is to blame? Maybe we are. If we have never taught them to express gratitude to others, how can we expect them to be grateful to us?

I know a man in Chicago who has cause to complain of the ingratitude of his stepsons. He slaved in a box factory, seldom earning more than forty dollars a week. He married a widow, and she persuaded him to borrow money and send her two grown sons to college. Out of his salary of forty dollars a week, he had to pay for food, rent, fuel, clothes, and also for the payments on his notes. He did this for four years, working like a coolie, and never complaining.

Did he get any thanks? No; his wife took it all for granted-and so did her sons. They never imagined that they owed their stepfather anything-not even thanks!

Who was to blame? The boys? Yes; but the mother was even more to blame. She thought it was a shame to burden their young lives with "a sense of obligation". She didn't want her sons to "start out under debt". So she never dreamed of saying: "What a prince your stepfather is to help you through college!" Instead, she took the attitude: "Oh, that's the least he can do."

She thought she was sparing her sons, but in reality, she was sending them out into life with the dangerous idea that the world owed them a living. And it was a dangerous idea- for one of those sons tried to "borrow" from an employer, and ended up in jail!

We must remember that our children are very much what we make them. For example, my mother's sister-Viola Alexander, of 144 West Minnehala Parkway, Minneapolis -is a shining example of a woman who has never had cause to complain about the "ingratitude" of children. When I was a boy, Aunt Viola took her own mother into her home to love and take care of; and she did the same thing for her husband's mother. I can still close my eyes and see those two old ladies sitting before the fire in Aunt Viola's farmhouse. Were they any "trouble" to Aunt Viola? Oh, often, I suppose. But you would never have guessed it from her attitude. She loved those old ladies-so she pampered them, and spoiled them, and made them feel at home. In addition, Aunt Viola had six children of her own; but it never occurred to her that she was doing anything especially noble, or deserved any halos for taking these old ladies into her home. To her, it was the natural thing, the right thing, the thing she wanted to do.

Where is Aunt Viola today? Well, she has now been a widow for twenty-odd years, and she has five grown-up children- five separate households-all clamouring to share her, and to have her come and live in their homes! Her children adore her; they never get enough of her. Out of "gratitude"? Nonsense! It is love-sheer love. Those children breathed in warmth and radiant human-kindness all during their childhoods. Is it any wonder that, now that the situation is reversed, they give back love?

So let us remember that to raise grateful children, we have to be grateful. Let us remember "little pitchers have big ears"-and watch what we say. To illustrate-the next time we are tempted to belittle someone's kindness in the presence of our children, let's stop. Let's never say: "Look at these dishcloths Cousin Sue sent for Christmas. She knit them herself. They didn't cost her a cent!" The remark may seem trivial to us-but the children are listening. So, instead, we had better say: "Look at the hours Cousin Sue spent making these for Christmas! Isn't she nice? Let's write her a thank-you note right now." And our children may unconsciously absorb the habit of praise and appreciation.

To avoid resentment and worry over ingratitude, here is Rule 3:

4.3

RULE 3

A. Instead of worrying about ingratitude, let's expect it. Let's remember that Jesus healed ten lepers in one day-and only one thanked him. Why should we expect more gratitude than Jesus got?

B. Let's remember that the only way to find happiness is not to expect gratitude, but to give for the joy of giving.

C. Let's remember that gratitude is a "cultivated" trait; so if we want our children to be grateful, we must train them to be grateful.

✣ ✣ ✣

Chapter 15

Would You Take A Million Dollars For What You Have?

I have known Harold Abbott for years. He lives at 820 South Madison Avenue, Webb City, Missouri. He used to be my lecture manager. One day he and I met in Kansas City and he drove me down to my farm at Belton, Missouri. During that drive, I asked him how he kept from worrying; and he told me an inspiring story that I shall never forget.

"I used to worry a lot," he said, "but one spring day in 1934, I was walking down West Dougherty Street in Webb City when I saw a sight that banished all my worries. It all happened in ten seconds, but during those ten seconds I learned more about how to live than I had learned in the previous ten years. For two years I had been running a grocery store in Webb City," Harold Abbott said, as he told me the story. "I had not only lost all my savings, but I had incurred debts that took me seven years to pay back. My grocery store had been closed the previous Saturday; and now I was going to the Merchants and Miners Bank to borrow money so I could go to Kansas City to look for a job.

I walked like a beaten man. I had lost all my fight and faith. Then suddenly I saw coming down the street a man who had no legs. He was sitting on a little wooden platform equipped with wheels from roller skates. He propelled himself along the street with a block of wood in each hand. I met him just after he had crossed the street and was starting to lift himself up a few inches over the kerb to the sidewalk. As he tilted his little wooden platform to an angle, his eyes met mine. He greeted me with a grand smile. 'Good morning, sir. It is a fine morning, isn't it?' he said with spirit. As I stood looking at him, I realised how rich I was. I had two legs. I could walk. I felt ashamed of my self-pity.

I said to myself if he can be happy, cheerful, and confident without legs, I certainly can with legs. I could already feel my chest lifting. I had intended to ask the Merchants and Miners Bank for only one hundred dollars. But now I had courage to ask for two hundred. I had intended to say that I wanted to go to Kansas City to try to get a job. But now I announced confidently that I wanted to go to Kansas City to get a job. I got the loan; and I got the job.

"I now have the following words pasted on my bathroom mirror, and I read them every morning as I shave:

> *I had the blues because I had no shoes,*
> *Until upon the street, I met a man who had no feet.*

I once asked Eddie Rickenbacker what was the biggest lesson he had learned from drifting about with his companions in life rafts for twenty-one days, hopelessly lost in the Pacific. "The biggest lesson I learned from that experience," he said, "was that if you have all the fresh water you want to drink and all the food you want to eat, you ought never to complain about anything."

Time ran an article about a sergeant who had been wounded on Guadalcanal. Hit in the throat by a shell fragment, this sergeant had had seven blood transfusions. Writing a note to his doctor, he asked: "Will I live?" The doctor replied: "Yes." He wrote another note, asking: "Will I be able to talk?" Again the answer was yes. He then wrote another note, saying: "Then what in hell am I worrying about?"

Why don't you stop right now and ask yourself: "What in the hell am I worrying about?" You will probably find that it is comparatively unimportant and insignificant.

About ninety per cent of the things in our lives are right and about ten per cent are wrong. If we want to be happy, all we have to do is to concentrate on the ninety per cent that are right and ignore the ten per cent that are wrong. If we want to be worried and bitter and have stomach ulcers, all we have to do is to concentrate on the ten per cent that are wrong and ignore the ninety per cent that are glorious.

The words "Think and Thank" are inscribed in many of the Cromwellian churches of England. These words ought to be inscribed in our hearts, too: "Think and Thank". Think of all we have to be grateful for, and thank God for all our boons and bounties.

Jonathan Swift, author of *Gulliver's Travels*, was the most devastating pessimist in English literature. He was so sorry that he had been born that he wore black and fasted on his birthdays; yet, in his despair, this supreme pessimist of English literature praised the great health-giving powers of cheerfulness and happiness. "The best doctors in the world," he declared, "are Doctor Diet, Doctor Quiet, and Doctor Merryman."

You and I may have the services of "Doctor Merryman" free every hour of the day by keeping our attention fixed on all the incredible riches we possess-riches exceeding by far the fabled treasures of Ali Baba. Would you sell both your eyes for a billion dollars? What would you take for your two legs? Your hands? Your hearing? Your children? Your family? Add up your assets, and you will find that you won't sell what you have for all the gold ever amassed by the Rockefellers, the Fords and the Morgans combined.

But do we appreciate all this? Ah, no. As Schopenhauer said: "We seldom think of what we have but always of what we lack." Yes, the tendency to "seldom think of what we have but always of what we lack" is the greatest tragedy on earth. It has probably caused more misery than all the wars and diseases in history.

It caused John Palmer to turn "from a regular guy into an old grouch", and almost wrecked his home. I know because he told me so.

Mr. Palmer lives at 30 19th Avenue, Paterson, New Jersey. "Shortly after I returned from the Army," he said, "I started in business for myself. I worked hard day and night. Things were going nicely. Then trouble started. I couldn't get parts and materials. I was afraid I would have to give up my business. I worried so much that I changed from a regular guy into an old grouch. I became so sour and cross that, well I didn't know it then; but I now realise that I came very near to losing my happy home. Then one day a young, disabled veteran who works for me said: 'Johnny, you ought to be ashamed of yourself. You take on as if you were the only person in the world with troubles. Suppose you do have to shut up shop for a while-so what? You can start up again when things get normal. You've got a lot to be thankful for. Yet you are always growling. Boy, how I wish I were in your shoes. I Look at me. I've got only one arm, and half of my face is shot away, and yet I am not complaining. If you don't stop your growling and grumbling, you will lose not only your business, but also your health, your home, and your friends!'

"Those remarks stopped me dead in my tracks. They made me realise how well off I was. I resolved then and there that I would change and be my old self again-and I did."

A friend of mine, Lucile Blake, had to tremble on the edge of tragedy before she learned to be happy about what she had instead of worrying over what she lacked.

I met Lucile years ago, when we were both studying short-story writing in the Columbia University School of Journalism. Nine years ago, she got the shock of her life. She was living then in Tucson, Arizonia. She had-well, here is the story as she told it to me:

"I had been living in a whirl: studying the organ at the University of Arizona, conducting a speech clinic in town, and teaching a class in musical appreciation at the Desert Willow Ranch, where I was staying. I was going in for parties, dances, horseback rides under the stars. One morning I collapsed. My heart! 'You will have to lie in bed for a year of complete rest,' the doctor said. He didn't encourage me to believe I would ever be strong again.

"In bed for a year! To be an invalid-perhaps to die! I was terror-stricken! Why did all this have to happen to me? What had I done to deserve it? I wept and wailed. I was bitter and rebellious. But I did go to bed as the doctor advised. A neighbour of mine, Mr. Rudolf, an artist, said to me: 'You think now that spending a year in bed will be a tragedy. But it won't be. You will have time to think and get acquainted with yourself. You will make more spiritual growth in these next few months than you have made during all your previous life.' I became calmer, and tried to develop a new sense of values.

I read books of inspiration. One day I heard a radio commentator say: 'You can express only what is in your own consciousness.' I had heard words like these many times before, but now they reached down inside me and took root. I resolved to think only the thoughts I wanted to live by: thoughts of joy, happiness, health. I forced myself each morning, as soon as I awoke, to go over all the things I had to be grateful for. No pain. A lovely young daughter. My eyesight. My hearing. Lovely music on the radio. Time to read. Good food. Good friends. I was so cheerful and had so many visitors that the doctor put up a sign saying that only one visitor at a time would be allowed in my cabin-and only at certain hours.

"Nine years have passed since then, and I now lead a full, active life. I am deeply grateful now for that year I spent in bed. It was the most valuable and the happiest year I spent in Arizona. The habit I formed then of counting my blessings each morning still remains with me. It is one of my most precious possessions. I am ashamed to realise that I never really learned to live until I feared I was going to die."

My dear Lucile Blake, you may not realise it, but you learned the same lesson that Dr. Samuel Johnson learned two hundred years ago. "The habit of looking on the best side of every event," said Dr. Johnson, "is worth more than a thousand pounds a year."

Those words were uttered, mind you, not by a professional optimist, but by a man who had known anxiety, rags, and hunger for twenty years-and finally became one of the most eminent writers of his generation and the most celebrated conversationalist of all time.

Logan Pearsall Smith packed a lot of wisdom into a few words when he said: "There are two things to aim at in life: first, to get what you want; and, after that, to enjoy it. Only the wisest of mankind achieve the second."

Would you like to know how to make even dishwashing at the kitchen sink a thrilling experience? If so, read an inspiring book of incredible courage by Borghild Dahl. It is called *I Wanted to See.*

This book was written by a woman who was practically blind for half a century. "I had only one eye," she writes, "and it was so covered with dense scars that I had to do all my seeing through one small opening in the left of the eye. I could see a book only by holding it up close to my face and by straining my one eye as hard as I could to the left."

But she refused to be pitied, refused to be considered "different". As a child, she wanted to play hopscotch with other children, but she couldn't see the markings. So after the other children had gone home, she got down on the ground and crawled along with her eyes near to the marks. She memorised every bit of the ground where she and her friends played and soon became an expert at running games. She did her reading at home, holding a book of large print so close to her eyes that her eyelashes brushed the pages. She earned two college degrees: an A.B. from the University of Minnesota and a Master of Arts from Columbia University.

She started teaching in the tiny village of Twin Valley, Minnesota, and rose until she became professor of journalism and literature at Augustana College in Sioux Falls, South Dakota. She taught there for thirteen years, lecturing before women's clubs and giving radio talks about books and authors. "In the back of my mind," she writes, "there had always lurked a fear of total blindness. In order to overcome this, I had adopted a cheerful, almost hilarious, attitude towards life."

Then in 1943, when she was fifty-two years old, a miracle happened: an operation at the famous Mayo Clinic. She could now see forty times as well as she had ever been able to see before.

A new and exciting world of loveliness opened before her. She now found it thrilling even to wash dishes in the kitchen sink. "I begin to play with the white fluffy suds in the dish-pan," she writes. "I dip my hands into them and I pick up a ball of tiny soap bubbles. I hold them up against the light, and in each of them I can see the brilliant colours of a miniature rainbow."

As she looked through the window above the kitchen sink, she saw "the flapping grey-black wings of the sparrows flying through the thick, falling snow."

She found such ecstasy looking at the soap bubbles and sparrows that she closed her book with these words: " 'Dear Lord,' I whisper, 'Our Father in Heaven, I thank Thee. I thank Thee.' "

Imagine thanking God because you can wash dishes and see rainbows in bubbles and sparrows flying through the snow.

You and I ought to be ashamed of ourselves. All the days of our years we have been living in a fairyland of beauty, but we have been too blind to see, too satiated to enjoy.

If we want to stop worrying and start living. Rule 4 is:

4.4

RULE 4

Count your blessings - not your troubles!

✢ ✢ ✢

Chapter 16

Find Yourself And Be Yourself: Remember There Is No One Else on Earth Like You

I have a letter from Mrs. Edith Allred, of Mount Airy, North Carolina: "As a child, I was extremely sensitive and shy," she says in her letter. "I was always overweight and my cheeks made me look even fatter than I was. I had an old-fashioned mother who thought it was foolish to make clothes look pretty. She always said: 'Wide will wear while narrow will tear'; and she dressed me accordingly. I never went to parties; never had any fun; and when I went to school, I never joined the other children in outside activities, not even athletics. I was morbidly shy. I felt I was 'different' from everybody else, and entirely undesirable.

"When I grew up, I married a man who was several years my senior. But I didn't change. My in-laws were a poised and self-confident family. They were everything I should have been but simply was not. I tried my best to be like them, but I couldn't. Every attempt they made to draw me out of myself only drove me further into my shell. I became nervous and irritable. I avoided all friends. I got so bad I even dreaded the sound of the doorbell ringing! I was a failure. I knew it; and I was afraid my husband would find it out. So, whenever we were in public, I tried to be gay, and overacted my part. I knew I overacted; and I would be miserable for days afterwards. At last I became so unhappy that I could see no point in prolonging my existence. I began to think of suicide."

What happened to change this unhappy woman's life? Just a chance remark!

"A chance remark," Mrs. Allred continued, "transformed my whole life. My mother-in-law was talking one day of how she brought her children up, and she said: 'No matter what happened, I always insisted on their being themselves.' ... 'On being themselves.' ... That remark is what did it! In a flash, I realised I had brought all this misery on myself by trying to fit myself into a pattern to which I did not conform.

"I changed overnight! I started being myself. I tried to make a study of my own personality. Tried to find out what I was. I studied my strong points. I learned all I could about colours and styles, and dressed in a way that I felt was becoming to me. I reached out to make friends. I joined an organisation, a small one at first, and was petrified with fright when they put me on a programme. But each time I spoke, I gained a little courage. It took a long while-but today I have more happiness than I ever dreamed possible. In rearing my own children, I have always taught them the lesson I had to learn from such bitter experience: No matter what happens, always be yourself!"

This problem of being willing to be yourself is "as old as history," says Dr. James Gordon Gilkey, "and as universal as human life." This problem of being unwilling to be yourself is the hidden spring behind many neuroses and psychoses and complexes. Angelo Patri has written thirteen books and thousands of syndicated newspaper articles on the subject of child training, and he says: "Nobody is so miserable as he who longs to be somebody and something other than the person he is in body and mind."

This craving to be something you are not is especially rampant in Hollywood. Sam Wood, one of Hollywood's best-known directors, says the greatest headache he has with aspiring young actors is exactly this problem: to make them be themselves. They all want to be second-rate Lana Turners, or third-rate Clark Gables. "The public has already had that flavour," Sam Wood keeps telling them; "now it wants something else."

Before he started directing such pictures as Good-bye, Mr. Chips and For Whom the Bell Tolls, Sam Wood spent years in the real-estate business, developing sales personalities. He declares that the same principles apply in the business world as in the world of moving pictures. You won't get anywhere playing the ape. You can't be a parrot. "Experience has taught me," says Sam Wood, "that it is safest to drop, as quickly as possible, people who pretend to be what they aren't."

I recently asked Paul Boynton, employment director for the Socony-Vacuum Oil Company, what is the biggest mistake people make in applying for jobs. He ought to know: he has interviewed more than sixty thousand job seekers; and he has written a book entitled *Ways to Get a Job*. He replied: "The biggest mistake people make in applying for jobs is in not being themselves. Instead of taking their hair down and being completely frank, they often try to give you the answers they think you want." But it doesn't work, because nobody wants a phony. Nobody ever wants a counterfeit coin.

A certain daughter of a street-car conductor had to learn that lesson the hard way. She longed to be a singer. But her face was her misfortune. She had a large mouth and protruding buck teeth. When she first sang in public-in a New Jersey night-club-she tried to pull down her upper lip to cover her teeth. She tried to act "glamorous". The result? She made herself ridiculous. She was headed for failure. However, there was a man in this night-club who heard the girl sing and thought she had talent. "See here," he said bluntly, "I've been watching your performance and I know what it is you're trying to hide. You're ashamed of your teeth." The girl was embarrassed, but the man continued: "What of it? Is there any particular crime in having buck teeth? Don't try to hide them! Open your mouth, and the audience will love you when they see you're not ashamed. Besides," he said shrewdly, "those teeth you're trying to hide may make your fortune!"

Cass Daley took his advice and forgot about her teeth. From that time on, she thought only about her audience. She opened her mouth wide and sang with such gusto and enjoyment that she became a top star in movies and radio. Other comedians are now trying to copy her!

The renowned William James was speaking of men who had never found themselves when he declared that the average man develops only ten per cent of his latent mental abilities. "Compared to what we ought to be," he wrote, "we are only half awake. We are making use of only a small part of our physical and mental resources. Stating the thing broadly, the human individual thus lives far within his limits. He possesses powers of various sorts which he habitually fails to use."

You and I have such abilities, so let's not waste a second worrying because we are not like other people. You are something new in this world. Never before, since the beginning of time, has there ever been anybody exactly like you; and never again throughout all the ages to come will there ever be anybody exactly like you again. The new science of genetics informs us that you are what you are largely as a result of twenty-four chromosomes contributed by your father and twenty-four chromosomes contributed by your mother. These forty-eight chromosomes comprise everything that determines what you inherit. In each chromosome there may be, says Amran Sheinfeld, "anywhere from scores to hundreds of genes -with a single gene, in some cases, able to change the whole life of an individual." Truly, we are "fearfully and wonderfully" made.

Even after your mother and father met and mated, there was only one chance in 300,000 billion that the person who is specifically you would be born! In other words, if you had 300,000 billion brothers and sisters, they might have all been different from you. Is all this guesswork? No. It is a scientific fact. If you would like to read more about it, go to your public library and borrow a book entitled *You and Heredity* by Amran Scheinfeld.

I can talk with conviction about this subject of being yourself because I feel deeply about it. I know what I am talking about. I know from bitter and costly experience. To illustrate: when I first came to New York from the cornfields of Missouri, I enrolled in the American Academy of Dramatic Arts. I aspired to be an actor. I had what I thought was a brilliant idea, a short cut to success, an idea so simple, so foolproof, that I couldn't understand why thousands of ambitious people hadn't already discovered it. It was this: I would study how the famous actors of that day-John Drew, Walter Hampden, and Otis Skinner-got their effects. Then I would imitate the best point of each one of them and make myself into a shining, triumphant combination of all of them. How silly! How absurd! I had to waste years of my life imitating other people before it penetrated through my thick Missouri skull that I had to be myself, and that I couldn't possibly be anyone else.

That distressing experience ought to have taught me a lasting lesson. But it didn't. Not me. I was too dumb. I had to learn it all over again. Several years later, I set out to write what I hoped would be the best book on public speaking for businessmen that had ever been written. I had the same foolish idea about writing this book that I had formerly had about acting: I was going to borrow the ideas of a lot of other writers and put them all in one book-a book that would have everything. So I got scores of books on public speaking and spent a year incorporating their ideas into my manuscript. But it finally dawned on me once again that I was playing the fool. This hodgepodge of other men's ideas that I had written was so synthetic, so dull, that no businessman would ever plod through it. So I tossed a year's work into the wastebasket, and started all over again.

This time I said to myself: "You've got to be Dale Carnegie, with all his faults and limitations. You can't possibly be anybody else." So I quit trying to be a combination of other men, and rolled up my sleeves and did what I should have done in the first place: I wrote a textbook on public speaking out of my own experiences, observations, and convictions as a speaker and a teacher of speaking. I learned for all time, I hope-the lesson that Sir Walter Raleigh learned. (I am not talking about the Sir Walter who threw his coat in the mud for the Queen to step on. I am talking about the Sir Walter Raleigh who was professor of English literature at Oxford back in 1904.) "I can't write a book commensurate with Shakespeare," he said, "but I can write a book by me."

Be yourself. Act on the sage advice that Irving Berlin gave the late George Gershwin. When Berlin and Gershwin first met, Berlin was famous but Gershwin was a struggling young composer working for thirty-five dollars a week in Tin Pan Alley. Berlin, impressed by Gershwin's ability, offered Gershwin a job as his musical secretary at almost three times the salary he was then getting. "But don't take the job," Berlin advised. "If you do, you may develop into a second-rate Berlin. But if you insist on being yourself, some day you'll become a first-rate Gershwin."

Gershwin heeded that warning and slowly transformed himself into one of the significant American composer of his generation.

Charlie Chaplin, Will Rogers, Mary Margaret McBride, Gene Autry, and millions of others had to learn the lesson I am trying to hammer home in this chapter. They had to learn the hard way-just as I did.

When Charlie Chaplin first started making films, the director of the pictures insisted on Chaplin's imitating a popular German comedian of that day. Charlie Chaplin got nowhere until he acted himself. Bob Hope had a similar experience: spent years in a singing-and-dancing act-and got nowhere until he began to wisecrack and be himself. Will Rogers twirled a rope in vaudeville for years without saying a word. He got nowhere until he discovered his unique gift for humour and began to talk as he twirled his rope.

When Mary Margaret McBride first went on the air, she tried to be an Irish comedian and failed. When she tried to be just what she was-a plain country girl from Missouri-she became one of the most popular radio stars in New York.

When Gene Autry tried to get rid of his Texas accent and dressed like city boys and claimed he was from New York, people merely laughed behind his back. But when he started twanging his banjo and singing cowboy ballads, Gene Autry started out on a career that made him the world's most popular cowboy both in pictures and on the radio.

You are something new in this world. Be glad of it. Make the most of what nature gave you. In the last analysis, all art is autobiographical. You can sing only what you are. You can paint only what you are. You must be what your experiences, your environment, and your heredity have made you.

For better or for worse, you must cultivate your own little garden. For better or for worse, you must play your own little instrument in the orchestra of life.

As Emerson said in his essay on "Self-Reliance" : "There is a time in every man's education when he arrives at the conviction that envy is ignorance; that imitation is suicide; that he must take himself for better, for worse, as his portion; that though the wide universe is full of good, no kernel of nourishing corn can come to him but through his toil bestowed on that plot of ground which is given him to till. The power which resides in him is new in nature, and none but he knows what that is which he can do, nor does he know until he has tried."

That is the way Emerson said it. But here is the way a poet -the late Douglas Malloch-said it:

If you can't be a pine on the top of the hill.
Be a scrub in the valley-but be
The best little scrub by the side of the rill;
Be a bush, if you can't be a tree.

If you can't be a bush, be a bit of the grass.
And some highway happier make;
If you can't be a muskie, then just be a bass-
But the liveliest bass in the lake!

We can't all be captains, we've got to be crew.
There's something for all of us here.
There's big work to do and there's lesser to do
And the task we must do is the near.

If you can't be a highway, then just be a trail,
If you can't be the sun, be a star;
It isn't by the size that you win or you fail-
Be the best of whatever you are!

To cultivate a mental attitude that will bring us peace and freedom from worry, here is Rule 5:

4.5

RULE 5

Let's not imitate others. Let's find ourselves and be ourselves.

✦ ✦ ✦

Chapter 17

If You Have A Lemon, Make A Lemonade

While writing this book, I dropped in one day at the University of Chicago and asked the Chancellor, Robert Maynard Hutchins, how he kept from worrying. He replied: "I have always tried to follow a bit of advice given me by the late Julius Rosenwald, President of Sears, Roebuck and Company: 'When you have a lemon, make lemonade.'"

That is what a great educator does. But the fool does the exact opposite. If he finds that life has handed him a lemon, he gives up and says: "I'm beaten. It is fate. I haven't got a chance." Then he proceeds to rail against the world and indulge in an orgy of self-pity. But when the wise man is handed a lemon, he says: "What lesson can I learn from this misfortune? How can I improve my situation? How can I turn this lemon into a lemonade?"

After spending a lifetime studying people and their hidden reserves of power, the great psychologist, Alfred Adler, declared that one of the wonder-filled characteristics of human beings is "their power to turn a minus into a plus."

Here is an interesting and stimulating story of a woman I know who did just that. Her name is Thelma Thompson, and she lives at 100 Morningside Drive, New York City. "During the war," she said, as she told me of her experience, "during the war, my husband was stationed at an Army training camp near the Mojave Desert, in New Mexico. I went to live there in order to be near him. I hated the place. I loathed it. I had never before been so miserable. My husband was ordered out on maneuvers in the Mojave Desert, and I was left in a tiny shack alone. The heat was unbearable-125 degrees in the shade of a cactus. Not a soul to talk to but Mexicans and Indians, and they couldn't speak English. The wind blew incessantly, and all the food I ate, and the very air I breathed, were filled with sand, sand, sand!

"I was so utterly wretched, so sorry for myself, that I wrote to my parents. I told them I was giving up and coming back home. I said I couldn't stand it one minute longer. I would rather be in jail! My father answered my letter with just two lines-two lines that will always sing in my memory-two lines that completely altered my life:

Two men looked out from prison bars,
One saw the mud, the other saw stars.

"I read those two lines over and over. I was ashamed of myself. I made up my mind I would find out what was good in my present situation. I would look for the stars.

"I made friends with the natives, and their reaction amazed me. When I showed interest in their weaving and pottery, they gave me presents of their favourite pieces which they had refused to sell to tourists. I studied the fascinating forms of the cactus and the yuccas and the Joshua trees. I learned about prairie dogs, watched for the desert sunsets, and hunted for seashells that had been left there millions of years ago when the sands of the desert had been an ocean floor.

"What brought about this astonishing change in me? The Mojave Desert hadn't changed. The Indians hadn't changed. But I had. I had changed my attitude of mind. And by doing so, I transformed a wretched experience into the most exciting adventure of my life. I was stimulated and excited by this new world that I had discovered. I was so excited. I wrote a book about it-a novel that was published under the title *Bright Ramparts*. I had looked out of my self-created prison and found the stars."

Thelma Thompson, you discovered an old truth that the Greeks taught five hundred years before Christ was born: "The best things are the most difficult."

Harry Emerson Fosdick repeated it again in the twentieth century: "Happiness is not mostly pleasure; it is mostly victory." Yes, the victory that comes from a sense of achievement, of triumph, of turning our lemons into lemonades.

I once visited a happy farmer down in Florida who turned even a poison lemon into lemonade. When he first got this farm, he was discouraged. The land was so wretched he could neither grow fruit nor raise pigs. Nothing thrived there but scrub oaks and rattlesnakes. Then he got his idea. He would turn his liability into an asset: he would make the most of these rattlesnakes. To everyone's amazement, he started canning rattlesnake meat. When I stopped to visit him a few years ago, I found that tourists were pouring in to see his rattlesnake farm at the rate of twenty thousand a year. His business was thriving. I saw poison from the fangs of his rattlers being shipped to laboratories to make anti-venom toxin; I saw rattlesnake skins being sold at fancy prices to make women's shoes and handbags. I saw canned rattlesnake meat being shipped to customers all over the world. I bought a picture postcard of the place and mailed it at the local post office of the village, which had been re-christened "Rattlesnake, Florida", in honour of a man who had turned a poison lemon into a sweet lemonade.

As I have travelled up and down and back and forth across America time after time, it has been my privilege to meet dozens of men and women who have demonstrated "their power to turn a minus into a plus".

The late William Bolitho, author of Twelve Against the Gods, put it like this: "The most important thing in life is not to capitalise on your gains. Any fool can do that. The really important thing is to profit from your losses. That requires intelligence; and it makes the difference between a man of sense and a fool."

Bolitho uttered those words after he had lost a leg in a railway accident. But I know a man who lost both legs and turned his minus into a plus. His name is Ben Fortson. I met him in a hotel elevator in Atlanta, Georgia. As I stepped into the elevator,

I noticed this cheerful-looking man, who had both legs missing, sitting in a wheel-chair in a corner of the elevator. When the elevator stopped at his floor, he asked me pleasantly if I would step to one corner, so he could manage his chair better. "So sorry," he said, "to inconvenience you"-and a deep, heart-warming smile lighted his face as he said it.

When I left the elevator and went to my room, I could think of nothing but this cheerful cripple. So I hunted him up and asked him to tell me his story.

"It happened in 1929," he told me with a smile. "I had gone out to cut a load of hickory poles to stake the beans in my garden. I had loaded the poles on my Ford and started back home. Suddenly one pole slipped under the car and jammed the steering apparatus at the very moment I was making a sharp turn. The car shot over an embankment and hurled me against a tree. My spine was hurt. My legs were paralysed.

"I was twenty-four when that happened, and I have never taken a step since."

Twenty-four years old, and sentenced to a wheel-chair for the rest of his life! I asked him how he managed to take it so courageously, and he said: "I didn't." He said he raged and rebelled. He fumed about his fate. But as the years dragged on, he found that his rebellion wasn't getting him anything except bitterness. "I finally realised," he said, "that other people were kind and courteous to me. So the least I could do was to be kind and courteous to them."

I asked if he still felt, after all these years, that his accident had been a terrible misfortune, and he promptly said: "No." He said: "I'm almost glad now that it happened." He told me that after he got over the shock and resentment, he began to live in a different world. He began to read and developed a love for good literature. In fourteen years, he said, he had read at least fourteen hundred books; and those books had opened up new horizons for him and made his life richer than he ever thought possible. He began to listen to good music; and he is now thrilled by great symphonies that would have bored him before. But the biggest change was that he had time to think. "For the first time in my life," he said, "I was able to look at the world and get a real sense of values. I began to realise that most of the things I had been striving for before weren't worth-while at all."

As a result of his reading, he became interested in politics, studied public questions, made speeches from his wheel-chair! He got to know people and people got to know him. Today Ben Fortson-still in his wheel-chair-is Secretary of State for the State of Georgia!

During the last thirty-five years, I have been conducting adult-education classes in New York City, and I have discovered that one of the major regrets of many adults is that they never went to college. They seem to think that not having a college education is a great handicap. I know that this isn't necessarily true because I have known thousands of successful men who never went beyond high school. So I often tell these students the story of a man I knew who had never finished even grade school. He was brought up in blighting poverty. When his father died, his father's friends had to chip in to pay for the coffin in which he was buried. After his father's death, his mother worked in an umbrella factory ten hours a day and then brought piecework home and worked until eleven o'clock at night.

The boy brought up in these circumstances went in for amateur dramatics put on by a club in his church. He got such a thrill out of acting that he decided to take up public speaking. This led him into politics. By the time he reached thirty, he was elected to the New York State legislature. But he was woefully unprepared for such a responsibility. In fact, he told me that frankly he didn't know what it was all about. He studied the long, complicated bills that he was supposed to vote on-but, as far as he was concerned, those bills might as well have been written in the language of the Choctaw Indians. He was worried and bewildered when he was made a member of the committee on forests before he had ever set foot in a forest. He was worried and bewildered when he was made a member of the State Banking Commission before he had ever had a bank account. He himself told me that he was so discouraged that he would have resigned from the legislature if he hadn't been ashamed to admit defeat to his mother. In despair, he decided to study sixteen hours a day and turn his lemon of ignorance into a lemonade of knowledge. By doing that, he transformed himself from a local politician into a national figure and made himself so outstanding that The New York Times called him "the best-loved citizen of New York".

I am talking about Al Smith.

Ten years after Al Smith set out on his programme of political self-education, he was the greatest living authority on the government of New York State. He was elected Governor of New York for four terms-a record never attained by any other man. In 1928, he was the Democratic candidate for President. Six great universities-including Columbia and Harvard-conferred honorary degrees upon this man who had never gone beyond grade school.

Al Smith himself told me that none of these things would ever have come to pass if he hadn't worked hard sixteen hours a day to turn his minus into a plus.

Nietzsche's formula for the superior man was "not only to bear up under necessity but to love it".

The more I have studied the careers of men of achievement the more deeply I have been convinced that a surprisingly large number of them succeeded because they started out with handicaps that spurred them on to great endeavour and great rewards. As William James said: "Our infirmities help us unexpectedly."

Yes, it is highly probable that Milton wrote better poetry because he was blind and that Beethoven composed better music because he was deaf.

Helen Keller's brilliant career was inspired and made possible because of her blindness and deafness.

If Tchaikovsky had not been frustrated-and driven almost to suicide by his tragic marriage-if his own life had not been pathetic, he probably would never have been able to compose his immortal "Symphonie Pathetique".

If Dostoevsky and Tolstoy had not led tortured lives, they would probably never have been able to write their immortal novels.

"If I had not been so great an invalid," wrote the man who changed the scientific concept of life on earth, "if I had not been so great an invalid, I should not have done so much work as I have accomplished." That was Charles Darwin's confession that his infirmities had helped him unexpectedly.

The same day that Darwin was born in England another baby was born in a log cabin in the forests of Kentucky. He, too, was helped by his infirmities. His name was Lincoln- Abraham Lincoln. If he had been reared in an aristocratic family and had had a law degree from Harvard and a happy married life, he would probably never have found in the depths of his heart the haunting words that he immortalised at Gettysburg, nor the sacred poem that he spoke at his second inauguration-the most beautiful and noble phrases ever uttered by a ruler of men: "With malice toward none; with charity for all ..."

Harry Emerson Fosdick says in his book, *The Power to See it Through*: "There is a Scandinavian saying which some of us might well take as a rallying cry for our lives: 'The north wind made the Vikings.' Wherever did we get the idea that secure and pleasant living, the absence of difficulty, and the comfort of ease, ever of themselves made people either good or happy? Upon the contrary, people who pity themselves go on pitying themselves even when they are laid softly on a cushion, but always in history character and happiness have come to people in all sorts of circumstances, good, bad, and indifferent, when they shouldered their personal responsibility. So, repeatedly the north wind has made the Vikings."

Suppose we are so discouraged that we feel there is no hope of our ever being able to turn our lemons into lemonade-then here are two reasons why we ought to try, anyway-two reasons why we have everything to gain and nothing to lose.

Reason one: We may succeed.

Reason two: Even if we don't succeed, the mere attempt to turn our minus into a plus will cause us to look forward instead of backward; it will replace negative thoughts with positive thoughts; it will release creative energy and spur us to get so busy that we won't have either the time or the inclination to mourn over what is past and forever gone.

Once when Ole Bull, the world-famous violinist, was giving a concert in Paris, the A string on his violin suddenly snapped. But Ole Bull simply finished the melody on three strings. "That is life," says Harry Emerson Fosdick, "to have your A string snap and finish on three strings."

That is not only life. It is more than life. It is life triumphant!

If I had the power to do so, I would have these words of William Bolitho carved in eternal bronze and hung in every schoolhouse in the land:

"The most important thing in life is not to capitalize on your gains. Any fool can do that. The really important thing is to profit from your losses. That requires intelligence; and it makes the difference between a man of sense and a fool."

So, to cultivate a mental attitude that will bring us peace and happiness, let's do something about Rule 6:

4.6

RULE 6

When fate hands us a lemon, let's try to make a lemonade.

✦ ✦ ✦

Chapter 18

How To Cure Melancholy In Fourteen Days

When I started writing this book, I offered a two-hundred-dollar prize for the most helpful and inspiring true story on "How I Conquered Worry".

The three judges for this contest were: Eddie Rickenbacker, president, Eastern Air Lines; Dr. Stewart W. McClelland, president, Lincoln Memorial University; H. V. Kaltenborn, radio news analyst. However, we received two stories so superb that the judges found it impossible to choose between them. So we divided the prize. Here is one of the stories that tied for first prize-the story of C.R. Burton (who works for Whizzer Motor Sales of Missouri, Inc.), 1067 Commercial Street, Springfield, Missouri.

"I lost my mother when I was nine years old, and my father when I was twelve," Mr. Burton wrote me. "My father was killed, but my mother simply walked out of the house one day nineteen years ago; and I have never seen her since. Neither have I ever seen my two little sisters that she took with her. She never even wrote me a letter until after she had been gone seven years. My father was killed in an accident three years after Mother left. He and a partner bought a cafe in a small Missouri town; and while Father was away on a business trip, his partner sold the cafe for cash and skipped out. A friend wired Father to hurry back home; and in his hurry, Father was killed in a car accident at Salinas, Kansas. Two of my father's sisters, who were poor and old and sick took three of the children into their homes. Nobody wanted me and my little brother. We were left at the mercy of the town. We were haunted by the fear of being called orphans and treated as orphans. Our fears soon materialised, too.

I lived for a little while with a poor family in town. But times were hard and the head of the family lost his job, so they couldn't afford to feed me any longer. Then Mr. and Mrs. Loftin took me to live with them on their farm eleven miles from town. Mr. Loftin was seventy years old, and sick in bed with shingles. He told me I could stay there 'as long as I didn't lie, didn't steal, and did as I was told'. Those three orders became my Bible. I lived by them strictly. I started to school, but the first week found me at home, bawling like a baby. The other children picked on me and poked fun at my big nose and said I was dumb and called me an 'orphan brat'. I was hurt so badly that I wanted to fight them; but Mr. Loftin, the farmer who had taken me in, said to me: 'Always remember that it takes a bigger man to walk away from a fight than it does to stay and fight.' I didn't fight until one day a kid picked up some chicken manure from the schoolhouse yard and threw it in my face. I beat the hell out of him; and made a couple of friends. They said he had it coming to him.

"I was proud of a new cap that Mrs. Loftin had bought me. One day one of the big girls jerked it off my head and filled it with water and ruined it. She said she filled it with water so that 'the water would wet my thick skull and keep my popcorn brains from popping'.

"I never cried at school, but I used to bawl it out at home. Then one day Mrs. Loftin gave me some advice that did away with all troubles and worries and turned my enemies into friends. She said: 'Ralph, they won't tease you and call you an "orphan brat" any more if you will get interested in them and see how much you can do for them.' I took her advice. I studied hard; and I soon headed the class. I was never envied because I went out of my way to help them.

"I helped several of the boys write their themes and essays. I wrote complete debates for some of the boys. One lad was ashamed to let his folks know that I was helping him. So he used to tell his mother he was going possum hunting. Then he would come to Mr. Loftin's farm and tie his dogs up in the barn while I helped him with his lessons. I wrote book reviews for one lad and spent several evenings helping one of the girls on her maths.

"Death struck our neighbourhood. Two elderly farmers died and one woman was deserted by her husband. I was the only male in four families. I helped these widows for two years. On my way to and from school, I stopped at their farms, cut wood for them, milked their cows, and fed and watered their stock. I was now blessed instead of cursed. I was accepted as a friend by everyone. They showed their real feelings when I returned home from the Navy. More than two hundred farmers came to see me the first day I was home. Some of them drove as far as eighty miles, and their concern for me was really sincere. Because I have been busy and happy trying to help other people, I have few worries; and I haven't been called an 'orphan brat' now for thirteen years."

Hooray for C.R. Burton! He knows how to win friends! And he also knows how to conquer worry and enjoy life.

So did the late Dr. Frank Loope, of Seattle, Washington. He was an invalid for twenty-three years. Arthritis. Yet Stuart Whithouse of the Seattle Star wrote me, saying: "I interviewed Dr. Loope many times; and I have never known a man more unselfish or a man who got more out of life."

How did this bed-ridden invalid get so much out of life? I'll give you two guesses. Did he do it by complaining and criticising? No. ... By wallowing in self-pity and demanding that he be the centre of attention and everyone cater to him? No. ... Still wrong. He did it by adopting as his slogan the motto of the Prince of Wales: "Ich dien"-"I serve." He accumulated the names and addresses of other invalids and cheered both them and himself by writing happy, encouraging letters. In fact, he organised a letterwriting club for invalids and got them writing letters to one another. Finally, he formed a national organisation called the Shut-in Society.

As he lay in bed, he wrote an average of fourteen hundred letters a year and brought joy to thousands of invalids by getting radios and books for shut-ins.

What was the chief difference between Dr. Loope and a lot of other people? Just this: Dr. Loope had the inner glow of a man with a purpose, a mission. He had the joy of knowing that he was being used by an idea far nobler and more significant than himself, instead of being as Shaw put it: "a selfcentred, little clod of ailments and grievances complaining that the world would not devote itself to making him happy."

Here is the most astonishing statement that I ever read from the pen of a great psychiatrist. This statement was made by Alfred Adler. He used to say to his melancholia patients: "You can be cured in fourteen days if you follow this prescription. Try to think every day how you can please someone."

That statement sounds so incredible that I feel I ought to try to explain it by quoting a couple of pages from Dr. Adler's splendid book, *What Life Should Mean to You*. (*) (By the way, there is a book you ought to read.)

"Melancholia," says Adler in *What Life Should Mean to You*: "is like a long-continued rage and reproach against others, though for the purpose of gaining care, sympathy and support, the patient seems only to be dejected about his own guilt. A melancholiac's first memory is generally something like this: 'I remember I wanted to lie on the couch, but my brother was lying there. I cried so much that he had to leave.'

"Melancholiacs are often inclined to revenge themselves by committing suicide, and the doctor's first care is to avoid giving them an excuse for suicide. I myself try to relieve the whole tension by proposing to them, as the first rule in treatment, 'Never do anything you don't like.' This seems to be very modest, but I believe that it goes to the root of the whole trouble If a melancholiac is able to do anything he wants, whom can he accuse? What has he got to revenge himself for? 'If you want to go to the theatre,' I tell him, 'or to go on a holiday, do it. If you find on the way that you don't want to, stop it.' It is the best situation anyone could be in. It gives a satisfaction to his striving

* Allen & Unwin Ltd.

for superiority. He is like God and can do what he pleases. On the other hand, it does not fit very easily into his style of life. He wants to dominate and accuse others and if they agree with him there is no way of dominating them. This rule is a great relief and I have never had a suicide among my patients.

"Generally the patient replies: 'But there is nothing I like doing.' I have prepared for this answer, because I have heard it so often. 'Then refrain from doing anything you dislike,' I say. Sometimes, however, he will reply: 'I should like to stay in bed all day.' I know that, if I allow it, he will no longer want to do it. I know that, if I hinder him, he will start a war. I always agree.

"This is one rule. Another attacks their style of life more directly. I tell them: 'You can be cured in fourteen days if you follow this prescription. Try to think every day how you can please someone.' See what this means to them. They are occupied with the thought. 'How can I worry someone.' The answers are very interesting. Some say: 'This will be very easy for me. I have done it all my life.' They have never done it. I ask them to think it over. They do not think it over. I tell them: 'You can make use of all the time you spend when you are unable to go to sleep by thinking how you can please someone, and it will be a big step forward in your health.' When I see them next day, I ask them: 'Did you think over what I suggested?' They answer: 'Last night I went to sleep as soon as I got to bed.' All this must be done, of course, in a modest, friendly manner, without a hint of superiority.

"Others will answer: 'I could never do it. I am so worried.' I tell them: 'Don't stop worrying; but at the same time you can think now and then of others.' I want to direct their interest always towards their fellows. Many say: 'Why should I please others? Others do not try to please me.' 'You must think of your health,' I answer. The others will suffer later on.' It is extremely rare that I have found a patient who said: 'I have thought over what you suggested.' All my efforts are devoted towards increasing the social interest of the patient. I know that the real reason for his malady is his lack of co-operation and I want him to see it too. As soon as he can connect himself with his fellow men on an equal and cooperative footing, he is cured. ... The most important task imposed by religion has always been 'Love thy neighbour'. ... It is the individual who is not interested in his fellow man who has the greatest difficulties in life and provides the greatest injury to others. It is from among such individuals that all human failures spring.

... All that we demand of a human being, and the highest praise we can give him is that he should be a good fellow worker, a friend to all other men, and a true partner in love and marriage."

Dr. Adler urges us to do a good deed every day. And what is a good deed? "A good deed," said the prophet Mohammed, "is one that brings a smile of joy to the face of another."

Why will doing a good deed every day produce such astounding efforts on the doer? Because trying to please others will cause us to stop thinking of ourselves: the very thing that produces worry and fear and melancholia.

Mrs. William T. Moon, who operates the Moon Secretarial School, 521 Fifth Avenue, New York, didn't have to spend two weeks thinking how she could please someone in order to banish her melancholy. She went Alfred Adler one better-no, she went Adler thirteen better. She banished her melancholy, not in fourteen days, but in one day, by thinking how she could please a couple of orphans.

It happened like this: "In December, five years ago," said Mrs. Moon, "I was engulfed in a feeling of sorrow and self-pity. After several years of happy married life, I had lost my husband. As the Christmas holidays approached, my sadness deepened. I had never spent a Christmas alone in all my life; and I dreaded to see this Christmas come. Friends had invited me to spend Christmas with them. But I did not feel up to any gaiety. I knew I would be a wet blanket at any party. So, I refused their kind invitations. As Christmas Eve approached, I was more and more overwhelmed with self-pity. True, I should have been thankful for many things, as all of us have many things for which to be thankful. The day before Christmas, I left my office at three o'clock in the afternoon and started walking aimlessly up Fifth Avenue, hoping that I might banish my self-pity and melancholy. The avenue was jammed with gay and happy crowds-scenes that brought back memories of happy years that were gone.

I just couldn't bear the thought of going home to a lonely and empty apartment. I was bewildered. I didn't know what to do. I couldn't keep the tears back. After walking aimlessly for an hour or so, I found myself in front of a bus terminal. I remembered that my husband and I had often boarded an unknown bus for adventure, so I boarded the first bus I found at the station. After crossing the Hudson River and riding for some time, I heard the bus conductor say: 'Last stop, lady.' I got off. I didn't even know the name of the town. It was a quiet, peaceful little place. While waiting for the next bus home, I started walking up a residential street. As I passed a church, I heard the beautiful strains of 'Silent Night'. I went in. The church was empty except for the organist. I sat down unnoticed in one of the pews. The lights from the gaily decorated Christmas tree made the decorations seem like myriads of stars dancing in the moonbeams. The long-drawn cadences of the music-and the fact that I had forgotten to eat since morning-made me drowsy. I was weary and heavy-laden, so I drifted off to sleep.

"When I awoke, I didn't know where I was. I was terrified. I saw in front of me two small children who had apparently come in to see the Christmas tree. One, a little girl, was pointing at me and saying: 'I wonder if Santa Clause brought her'. These children were also frightened when I awoke. I told them that I wouldn't hurt them. They were poorly dressed. I asked them where their mother and daddy were. 'We ain't got no mother and daddy,' they said. Here were two little orphans much worse off than I had ever been. They made me feel ashamed of my sorrow and self-pity. I showed them the Christmas tree and then took them to a drugstore and we had some refreshments, and I bought them some candy and a few presents. My loneliness vanished as if by magic. These two orphans gave me the only real happiness and self-forgetfulness that I had had in months.

As I chatted with them, I realised how lucky I had been. I thanked God that all my Christmases as a child had been bright with parental love and tenderness. Those two little orphans did far more for me than I did for them. That experience showed me again the necessity of making other people happy in order to be happy ourselves. I found that happiness is contagious. By giving, we receive. By helping someone and giving out love, I had conquered worry and sorrow and self-pity, and felt like a new person. And I was a new person-not only then, but in the years that followed." I could fill a book with stories of people who forgot themselves into health and happiness. For example, let's take the case of Margaret Tayler Yates, one of the most popular women in the United States Navy.

Mrs. Yates is a writer of novels, but none of her mystery stories is half so interesting as the true story of what happened to her that fateful morning when the Japanese struck our fleet at Pearl Harbour. Mrs. Yates had been an invalid for more than a year: a bad heart. She spent twenty-two out of every twentyfour hours in bed. The longest journey that she undertook was a walk into the garden to take a sunbath. Even then, she had to lean on the maid's arm as she walked. She herself told me that in those days she expected to be an invalid for the balance of her life. "I would never have really lived again," she told me," if the Japs had not struck Pearl Harbour and jarred me out of my complacency.

"When this happened," Mrs. Yates said, as she told her story, "everything was chaos and confusion. One bomb struck so near my home, the concussion threw me out of bed. Army trucks rushed out to Hickam Field, Scofield Barracks, and Kaneohe Bay Air Station, to bring Army and Navy wives and children to the public schools. There the Red Cross telephoned those who had extra rooms to take them in. The Red Cross workers knew that I had a telephone beside my bed, so they asked me to be a clearing-house of information. So I kept track of where Army and Navy wives and children were being housed, and all Navy and Army men were instructed by the Red Cross to telephone me to find out where their families were.

"I soon discovered that my husband, Commander Robert Raleigh Yates, was safe. I tried to cheer up the wives who did not know whether their husbands had been killed; and I tried to give consolation to the widows whose husbands had been killed-and they were many. Two thousand, one hundred and seventeen officers and enlisted men in the Navy and Marine Corps were killed and 960 were reported missing.

"At first I answered these phone calls while lying in bed. Then I answered them sitting up in bed. Finally, I got so busy, so excited, that I forgot all about my weakness and got out of bed and sat by a table. By helping others who were much worse off than I was, I forgot all about myself; and I have never gone back to bed again except for my regular eight hours of sleep each night. I realise now that if the Japs had not struck at Pearl Harbour, I would probably have remained a semi-invalid all my life. I was comfortable in bed. I was constantly waited on, and I now realise that I was unconsciously losing my will to rehabilitate myself.

"The attack on Pearl Harbour was one of the greatest tragedies in American history, but as far as I was concerned, it was one of the best things that ever happened to me. That terrible crisis gave me strength that I never dreamed I possessed. It took my attention off myself and focused it on others. It gave me something big and vital and important to live for. I no longer had time to think about myself or care about myself."

A third of the people who rush to psychiatrists for help could probably cure themselves if they would only do as Margaret Yates did: get interested in helping others. My idea? No, that is approximately what Carl Jung said. And he ought to know -if anybody does. He said: "About one-third of my patients are suffering from no clinically definable neurosis, but from the senselessness and emptiness of their lives." To put it another way, they are trying to thumb a ride through life-and the parade passes them by. So they rush to a psychiatrist with their petty, senseless, useless lives. Having missed the boat, they stand on the wharf, blaming everyone except themselves and demanding that the world cater to their self-centred desires.

You may be saying to yourself now: "Well, I am not impressed by these stories. I myself could get interested in a couple of orphans I met on Christmas Eve; and if I had been at Pearl Harbour, I would gladly have done what Margaret Tayler Yates did. But with me things are different: I live an ordinary humdrum life. I work at a dull job eight hours a day. Nothing dramatic ever happens to me. How can I get interested in helping others? And why should I? What is there in it for me?"

A fair question. I'll try to answer it. However humdrum your existence may be, you surely meet some people every day of your life. What do you do about them? Do you merely stare through them, or do you try to find out what it is that makes them tick? How about the postman, for example-he walks hundreds of miles every year, delivering mail to your door; but have you ever taken the trouble to find out where he lives, or ask to see a snapshot of his wife and his kids? Did you ever ask him if his feet get tired, or if he ever gets bored?

What about the grocery boy, the newspaper vendor, the chap at the corner who polishes your shoes? These people are human -bursting with troubles, and dreams, and private ambitions. They are also bursting for the chance to share them with someone. But do you ever let them? Do you ever show an eager, honest interest in them or their lives? That's the sort of thing I mean. You don't have to become a Florence Nightingale or a social reformer to help improve the world-your own private world; you can start tomorrow morning with the people you meet!

What's in it for you? Much greater happiness! Greater satisfaction, and pride in yourself! Aristotle called this kind of attitude "enlightened selfishness". Zoroaster said: "Doing good to others is not a duty. It is a joy, for it increases your own health and happiness." And Benjamin Franklin summed it up very simply-"When you are good to others," said Franklin, "you are best to yourself."

"No discovery of modern psychology," writes Henry C. Link, director of the Psychological Service Centre in New York, "no discovery of modern psychology is, in my opinion, so important as its scientific proof of the necessity of self-sacrifice or discipline to self-realisation and happiness."

Thinking of others will not only keep you from worrying about yourself; it will also help you to make a lot of friends and have a lot of fun. How? Well, I once asked Professor William Lyon Phelps, of Yale, how he did it; and here is what he said:

"I never go into a hotel or a barber-shop or a store without saying something agreeable to everyone I meet. I try to say something that treats them as an individual-not merely a cog in a machine. I sometimes compliment the girl who waits on me in the store by telling her how beautiful her eyes are - or her hair. I will ask a barber if he doesn't get tired standing on his feet all day. I'll ask him how he came to take up barbering- how long he has been at it and how many heads of hair he has cut. I'll help him figure it out. I find that taking an interest in people makes them beam with pleasure. I frequently shake hands with a redcap who has carried my grip. It gives him a new lift and freshens him up for the whole day. One extremely hot summer day, I went into the dining car of the New Haven Railway to have lunch. The crowded car was almost like a furnace and the service was slow.

When the steward finally got around to handing me the menu, I said: 'The boys back there cooking in that hot kitchen certainly must be suffering today.' The steward began to curse. His tones were bitter. At first, I thought he was angry. 'Good God Almighty,' he exclaimed, 'the people come in here and complain about the food. They kick about the slow service and growl about the heat and the prices. I have listened to their criticisms for nineteen years and you are the first person and the only person that has ever expressed any sympathy for the cooks back there in the boiling kitchen. I wish to God we had more passengers like you.'

"The steward was astounded because I had thought of the coloured cooks as human beings, and not merely as cogs in the organisation of a great railway. What people want," continued Professor Phelps, "is a little attention as human beings. When I meet a man on the street with a beautiful dog, I always comment on the dog's beauty. As I walk on and glance back over my shoulder, I frequently see the man petting and admiring the dog. My appreciation has renewed his appreciation. "One time in England, I met a shepherd, and expressed my sincere admiration for his big intelligent sheepdog. I asked him to tell me how he trained the dog. As I walked away, I glanced back over my shoulder and saw the dog standing with his paws on the shepherd's shoulders and the shepherd was petting him. By taking a little interest in the shepherd and his dog, I made the shepherd happy. I made the dog happy and I made myself happy."

Can you imagine a man who goes around shaking hands with porters and expressing sympathy for the cooks in the hot kitchen-and telling people how much he admires their dogs- can you imagine a man like that being sour and worried and needing the services of a psychiatrist? You can't, can you? No, of course not. A Chinese proverb puts it this way: "A bit of fragrance always clings to the hand that gives you roses."

You didn't have to tell that to Billy Phelps of Yale. He knew it. He lived it.

If you are a man, skip this paragraph. It won't interest you. It tells how a worried, unhappy girl got several men to propose to her. The girl who did that is a grandmother now. A few years ago, I spent the night in her and her husband's home. I had been giving a lecture in her town; and the next morning she drove me about fifty miles to catch a train on the main line to New York Central. We got to talking about winning friends, and she said: "Mr. Carnegie, I am going to tell you something that I have never confessed to anyone before- not even to my husband." (By the way, this story isn't going to be half so interesting as you probably imagine.) She told me that she had been reared in a social-register family in Philadelphia. "The tragedy of my girlhood and young womanhood," she said, "was our poverty. We could never entertain the way the other girls in my social set entertained.

My clothes were never of the best quality. I outgrew them and they didn't fit and they were often out of style. I was so humiliated, so ashamed, that I often cried myself to sleep. Finally, in sheer desperation, I hit upon the idea of always asking my partner at dinner-parties to tell me about his experiences, his ideas, and his plans for the future. I didn't ask these questions because I was especially interested in the answers. I did it solely to keep my partner from looking at my poor clothes. But a strange thing happened: as I listened to these young men talk and learned more about them, I really became interested in listening to what they had to say. I became so interested that I myself sometimes forgot about my clothes. But the astounding thing to me was this: since I was a good listener and encouraged the boys to talk about themselves, I gave them happiness and I gradually became the most popular girl in our social group and three of these men proposed marriage to me."

(There you are, girls: that is the way it is done.)

Some people who read this chapter are going to say: "All this talk about getting interested in others is a lot of damn nonsense! Sheer religious pap! None of that stuff for me! I am going to put money in my purse. I am going to grab all I can get-and grab it now-and to hell with the other dumb clucks!"

Well, if that is your opinion, you are entitled to it; but if you are right, then all the great philosophers and teachers since the beginning of recorded history-Jesus, Confucius, Buddha, Plato, Aristotle, Socrates, Saint Francis-were all wrong. But since you may sneer at the teachings of religious leaders, let's turn for advice to a couple of atheists. First, let's take the late A. E. Housman, professor at Cambridge University, and one of the most distinguished scholars of his generation. In 1936, he gave an address at Cambridge University on "The Name and Nature of Poetry." It is that address, he declared that "the greatest truth ever uttered and the most profound moral discovery of all time were those words of Jesus: 'He that findeth his life shall lose it: and he that loseth his life for my sake shall find it.'"

We have heard preachers say that all our lives. But Housman was an atheist, a pessimist, a man who contemplated suicide; and yet he felt that the man who thought only of himself wouldn't get much out of life. He would be miserable. But the man who forgot himself in service to others would find the joy of living.

If you are not impressed by what A.E. Housman said, let's turn for advice to the most distinguished American atheist of the twentieth century: Theodore Dreiser. Dreiser ridiculed all religions as fairy tales and regarded life as "a tale told by an idiot, full of sound and fury, signifying nothing." Yet Dreiser advocated the one great principle that Jesus taught- service to others. "If he [man] is to extract any joy out of his span," Dreiser said, "he must think and plan to make things better not only for himself but for others, since joy for himself depends upon his joy in others and theirs in him." If we are going "to make things better for others"-as Dreiser advocated-let's be quick about it. Time is a-wastin'. "I shall pass this way but once. Therefore any good that I can do or any kindness that I can show-let me do it now. Let me not defer nor neglect it, for I shall not pass this way again."

So if you want to banish worry and cultivate peace and happiness, here is Rule 7:

> 4.7
>
> RULE 7
>
> Forget yourself by becoming interested in others. Do every day a good deed that will put a smile of joy on someone's face.

Part Four In A Nutshell - Seven Ways To Cultivate A Mental Attitude That Will Bring You Peace And Happiness

RULE 1 Let's fill our minds with thoughts of peace, courage, health, and hope, for "our life is what our thoughts make it".

RULE 2 Let's never try to get even with our enemies, because if we do we will hurt ourselves far more than we hurt them. Let's do as General Eisenhower does: let's never waste a minute thinking about people we don't like.

RULE 3 A. Instead of worrying about ingratitude, let's expect it. Let's remember that Jesus healed ten lepers in one day-and only one thanked Him. Why should we expect more gratitude than Jesus got?

B. Let's remember that the only way to find happiness is not to expect gratitude-but to give for the joy of giving.

C. Let's remember that gratitude is a "cultivated" trait; so if we want our children to be grateful, we must train them to be grateful.

RULE 4 Count your blessings-not your troubles!

RULE 5 Let's not imitate others. Let's find ourselves and be ourselves, for "envy is ignorance" and "imitation is suicide".

RULE 6 When fate hands us a lemon, let's try to make a lemonade.

RULE 7 Let's forget our own unhappiness-by trying to create a little happiness for others. "When you are good to others, you are best to yourself.

PART FIVE

The Golden Rule For Conquering Worry

Chapter 19

How My Mother And Father Conquered Worry

As I have said, I was born and brought up on a Missouri farm. Like most farmers of that day, my parents had pretty hard scratching. My mother had been a country schoolteacher and my father had been a farm hand working for twelve dollars a month. Mother made not only my clothes, but also the soap with which we washed our clothes.

We rarely had any cash-except once a year when we sold our hogs. We traded our butter and eggs at the grocery store for flour, sugar, coffee. When I was twelve years old, I didn't have as much as fifty cents a year to spend on myself. I can still remember the day we went to a Fourth-of-July celebration and Father gave me ten cents to spend as I wished. I felt the wealth of the Indies was mine.

I walked a mile to attend a one-room country school. I walked when the snow was deep and the thermometer shivered around twenty-eight degrees below zero. Until I was fourteen, I never had any rubbers or overshoes. During the long, cold winters, my feet were always wet and cold. As a child I never dreamed that anyone had dry, warm feet during the winter.

My parents slaved sixteen hours a day, yet we constantly were oppressed by debts and harassed by hard luck. One of my earliest memories is watching the flood waters of the 102 River rolling over our corn- and hayfields, destroying everything. The floods destroyed our crops six years out of seven. Year after year, our hogs died of cholera and we burned them. I can close my eyes now and recall the pungent odour of burning hog flesh.

One year, the floods didn't come. We raised a bumper corn crop, bought feed cattle, and fattened them with our corn. But the floods might just as well have drowned our corn that year, for the price of fat cattle fell on the Chicago market; and after feeding and fattening the cattle, we got only thirty dollars more for them than what we had paid for them. Thirty dollars for a whole year's work!

No matter what we did, we lost money. I can still remember the mule colts that my father bought. We fed them for three years, hired men to break them, then shipped them to Memphis, Tennessee-and sold them for less than what we had paid for them three years previously.

After ten years of hard, grueling work, we were not only penniless; we were heavily in debt. Our farm was mortgaged. Try as hard as we might, we couldn't even pay the interest on the mortgage. The bank that held the mortgage abused and insulted my father and threatened to take his farm away from him. Father was forty-seven years old. After more than thirty years of hard work, he had nothing but debts and humiliation. It was more than he could take. He worried. His health broke. He had no desire for food; in spite of the hard physical work he was doing in the field all day, he had to take medicine to give him an appetite. He lost flesh. The doctor told my mother that he would be dead within six months. Father was so worried that he no longer wanted to live. I have often heard my mother say that when Father went to the barn to feed the horses and milk the cows, and didn't come back as soon as she expected, she would go out to the barn, fearing that she would find his body dangling from the end of a rope. One day as he returned home from Maryville, where the banker had threatened to foreclose the mortgage, he stopped his horses on a bridge crossing the 102 River, got off the wagon, and stood for a long time looking down at the water, debating with himself whether he should jump in and end it all.

Years later, Father told me that the only reason he didn't jump was because of my mother's deep, abiding, and joyous belief that if we loved God and kept His commandments everything would come out all right. Mother was right. Everything did come out all right in the end. Father lived forty-two happy years longer, and died in 1941, at the age of eighty-nine.

During all those years of struggle and heartache, my mother never worried. She took all her troubles to God in prayer. Every night before we went to bed, Mother would read a chapter from the Bible; frequently Mother or Father would read these comforting words of Jesus: "In my Father's house are many mansions. ... I go to prepare a place for you ... that where I am, there ye may be also." Then we all knelt down before our chairs in that lonely Missouri farmhouse and prayed for God's love and protection.

When William James was professor of philosophy at Harvard, he said: "Of course, the sovereign cure for worry is religious faith."

You don't have to go to Harvard to discover that. My mother found that out on a Missouri farm. Neither floods nor debts nor disaster could suppress her happy, radiant, and victorious spirit. I can still hear her singing as she worked:

Peace, peace, wonderful peace,
Flowing down from the Father above,
Sweep over my spirit forever I pray
In fathomless billows of love.

My mother wanted me to devote my life to religious work. I thought seriously of becoming a foreign missionary. Then I went away to college; and gradually, as the years passed, a change came over me. I studied biology, science, philosophy, and comparative religions. I read books on how the Bible was written. I began to question many of its assertions. I began to doubt many of the narrow doctrines taught by the country preachers of that day. I was bewildered. Like Walt Whitman, I "felt curious, abrupt questionings stir within me". I didn't know what to believe. I saw no purpose in life. I stopped praying. I became an agnostic.

I believed that all life was planless and aimless. I believed that human beings had no more divine purpose than had the dinosaurs that roamed the earth two hundred million years ago. I felt that some day the human race would perish-just as the dinosaurs had. I knew that science taught that the sun was slowly cooling and that when its temperature fell even ten per cent, no form of life could exist on earth. I sneered at the idea of a beneficent God who had created man in His own likeness. I believed that the billions upon billions of suns whirling through black, cold, lifeless space had been created by blind force. Maybe they had never been created at all. Maybe they existed forever-just as time and space have always existed.

Do I profess to know the answers to all these questions now? No. No man has ever been able to explain the mystery of the universe-the mystery of life. We are surrounded by mysteries. The operation of your body is a profound mystery. So is the electricity in your home. So is the flower in the crannied wall. So is the green grass outside your window. Charles F. Kettering, the guiding genius of General Motors Research Laboratories, has been giving Antioch College thirty thousand dollars a year out of his own pocket to try to discover why grass is green. He declares that if we knew how grass is able to transform sunlight, water, and carbon dioxide into food sugar, we could transform civilisation.

Even the operation of the engine in your car is a profound mystery. General Motors Research Laboratories have spent years of time and millions of dollars trying to find out how and why a spark in the cylinder sets off an explosion that makes your car run; and they don't know the answer.

The fact that we don't understand the mysteries of our bodies or electricity or a gas engine doesn't keep us from using and enjoying them. The fact that I don't understand the mysteries of prayer and religion no longer keeps me from enjoying the richer, happier life that religion brings. At long last, I realise the wisdom of Santayana's words: "Man is not made to understand life, but to live it."

I have gone back-well, I was about to say that I had gone back to religion; but that would not be accurate. I have gone forward to a new concept of religion. I no longer have the faintest interest in the differences in creeds that divide the Churches. But I am tremendously interested in what religion does for me, just as I am interested in what electricity and good food and water do for me. They help me to lead a richer, fuller, happier life. But religion does far more than that. It brings me spiritual values. It gives me, as William James puts it, "a new zest for life ... more life, a larger, richer, more satisfying life." It gives me faith, hope, and courage. It banishes tensions, anxieties, fears, and worries. It gives purpose to my life-and direction. It vastly improves my happiness. It gives me abounding health. It helps me to create for myself "an oasis of peace amidst the whirling sands of life".

Francis Bacon was right when he said, three hundred and fifty years ago: "A little philosophy inclineth man's mind to atheism; but depth in philosophy bringeth men's minds about to religion."

I can remember the days when people talked about the conflict between science and religion. But no more. The newest of all sciences-psychiatry-is teaching what Jesus taught. Why? Because psychiatrists realise that prayer and a strong religious faith will banish the worries, the anxieties, the strains and fears that cause more than half of all our ills. They know, as one of their leaders, Dr. A. A. Brill said: "Anyone who is truly religious does not develop a neurosis."

If religion isn't true, then life is meaningless. It is a tragic farce.

I interviewed Henry Ford a few years prior to his death. Before I met him, I had expected him to show the strains of the long years he had spent in building up and managing one of the world's greatest businesses. So I was surprised to how calm and well and peaceful he looked at seventy-eight. When I asked him if he ever worried, he replied: "No. I believe God is managing affairs and that He doesn't need any advice from me. With God in charge, I believe that every-thing will work out for the best in the end. So what is there to worry about?"

Today, even psychiatrists are becoming modern evangelists. They are not urging us to lead religious lives to avoid hell-fires in the next world, but they are urging us to lead religious lives to avoid the hell-fires of this world-the hell-fires of stomach ulcer, angina pectoris, nervous breakdowns, and insanity. As an example of what our psychologists and psychiatrists are teaching, read *The Return to Religion*, by Dr. Henry C. Link. You will probably find a copy in your public library.

Yes, the Christian religion is an inspiring, health-giving activity. Jesus said: "I came that ye might have life and have it more abundantly." Jesus denounced and attacked the dry forms and dead rituals that passed for religion in His day. He was a rebel. He preached a new kind of religion-a religion that threatened to upset the world. That is why He was crucified. He preached that religion should exist for man- not man for religion; that the Sabbath was made for man- not man for the Sabbath. He talked more about fear than He did about sin. The wrong kind of fear is a sin-a sin against your health, a sin against the richer, fuller, happier, courageous life that Jesus advocated. Emerson spoke of himself as a "Professor of the Science of Joy". Jesus, too, was a teacher of "the Science of Joy". He commanded His disciples to "rejoice and leap for joy".

Jesus declared that there were only two important things about religion: loving God with all our heart, and our neighbour as ourselves. Any man who does that is religious, regardless of whether he knows it. For example, my father-in-law, Henry Price, of Tulsa, Oklahoma. He tries to live by the golden rule; and he is incapable of doing anything mean, selfish, or dishonest. However, he doesn't attend church, and regards himself as an agnostic. Nonsense! What makes a man a Christian? I'll let John Baillie answer that. He was probably the most distinguished professor who ever taught theology at the University of Edinburgh. He said: "What makes a man a Christian is neither his intellectual acceptance of certain ideas, nor his conformity to a certain rule, but his possession of a certain Spirit, and his participation in a certain Life."

If that makes a man a Christian, then Henry Price is a noble one.

William James-the father of modern psychology-wrote to his friend, Professor Thomas Davidson, saying that as the years went by, he found himself "less and less able to get along without God".

Earlier in this book I mentioned that when the judges tried to pick the best story on worry sent in by my students, they had so much difficulty in choosing between two outstanding stories that the prize money was split. Here is the second story that tied for first prize-the unforgettable experience of a woman who had to find out the hard way that "she couldn't get along without God".

I am calling this woman Mary Cushman, although that is not her actual name. She has children and grandchildren who might be embarrassed to see her story in print, so I agreed to disguise her identity. However, the woman herself is real- very real. A few months ago, she sat in the armchair beside my desk and told me her story. Here is how it goes:

"During the depression," she said, "my husband's average salary was eighteen dollars a week. Many times we didn't have even that because he didn't get paid when he was ill-and that was often. He had a series of minor accidents; he also had mumps, scarlet fever, and repeated attacks of flu. We lost the little house that we had built with our own hands. We owed fifty dollars at the grocery store-and had five children to feed. I took in washing and ironing from the neighbours, and bought second-hand clothes from the Salvation Army store and made them over for my children to wear. I made myself ill with worry. One day the grocer to whom we owed fifty dollars accused my eleven-year-old boy of stealing a couple of pencils.

My son wept as he told me about it. I knew he was honest and sensitive-and I knew that he had been disgraced and humiliated in front of other people. That was the straw that broke my back. I thought of all the misery we had endured; and I couldn't see any hope for the future. I must have become temporarily insane with worry, for I shut off my washing machine, took my little five-year-old daughter into the bedroom, and plugged up the windows and cracks with paper and rags. My little girl said to me: 'Mommy, what are you doing?' and I answered: 'There's a little draught in here.' Then I turned on the gas heater we had in the bedroom-and didn't light it. As I lay down on the bed with my daughter beside me, she said: 'Mommy, this is funny-we just got up a little while ago!' But I said: 'Never mind, we'll take a little nap.'

Then I closed my eyes, listening to the gas escape from the heater. I shall never forget the smell of that gas. ...

"Suddenly I thought I heard music. I listened. I had forgotten to turn the radio off in the kitchen. It didn't matter now. But the music kept on, and presently I heard someone singing an old hymn:

> What a Friend we have in Jesus,
> All our sins and grief's to bear!
> What a privilege to carry
> Everything to God in prayer.
> Oh, what peace we often forfeit
> Oh, what needless pain we bear
> All because we do not carry
> Everything to God in prayer!

"As I listened to that hymn, I realised that I had made a tragic mistake. I had tried to fight all my terrible battles alone. I had not taken everything to God in prayer. ... I jumped up, turned off the gas, opened the door, and raised the windows.

"I wept and prayed all the rest of that day. Only I didn't pray for help-instead I poured out my soul in thanksgiving to God for the blessings He had given me: five splendid children- all of them healthy and fine, strong in body and mind. I promised God that never again would I prove so ungrateful. And I have kept that promise.

"Even after we lost our home, and had to move into a little country schoolhouse that we rented for five dollars a month, I thanked God for that schoolhouse; I thanked Him for the fact that I at least had a roof to keep us warm and dry. I thanked God honestly that things were not worse-and I believe that He heard me. For in time things improved-oh, not overnight; but as the depression lightened, we made a little more money. I got a job as a hat-check girl in a large country club, and sold stockings as a side line. To help put himself through college, one of my sons got a job on a farm, milked thirteen cows morning and night. Today my children are grown up and married; I have three fine grandchildren. And, as I look back on that terrible day when I turned on the gas, I thank God over and over that I 'woke up' in time. What joys I would have missed if I had carried out that act! How many wonderful years I would have forfeited forever! Whenever I hear now of someone who wants to end his life, I feel like crying out: 'Don't do it! Don't!' The blackest moments we live through can only last a little time-and then comes the future. ..."

On the average, someone commits suicide in the United States every thirty-five minutes. On the average, someone goes insane every hundred and twenty seconds. Most of these suicides-and probably many of the tragedies of insanity- could have been prevented if these people had only had the solace and peace that are found in religion and prayer.

One of the most distinguished psychiatrists living, Dr. Carl Jung, says in his book *Modern Man in Search of a Soul* (*):

"During the past thirty years, people from all the civilised countries of the earth have consulted me. I have treated many hundreds of patients. Among all my patients in the second half of life-that is to say, over thirty-five-there has not been one whose problem in the last resort was not that of finding a religious outlook on life. It is safe to say that every one of them fell ill because he had lost that which the living religions of every age have given to their followers, and none of them has been really healed who did not regain his religious outlook."

That statement is so significant I want to repeat it in bold type.

Dr. Carl Jung said:

"During the past thirty years, people from all the civilised countries of the earth have consulted me. I have treated many hundreds of patients. Among all my patients in the second half

---

* Kegar Paul, Trench, Trubner & Co. Ltd.

of life - that is to say, over thirty-five-there has not been one whose problem in the last resort was not that of finding a religious outlook on life. It is safe to say that every one of them fell ill because he had lost that which the living religions of every age have given to their followers, and none of them has been really healed who did not regain his religious outlook."

William James said approximately the same thing: "Faith is one of the forces by which men live," he declared, "and the total absence of it means collapse."

The late Mahatma Gandhi, the greatest Indian leader since Buddha, would have collapsed if he had not been inspired by the sustaining power of prayer. How do I know? Because Gandhi himself said so. "Without prayer," he wrote, "I should have been a lunatic long ago."

Thousands of people could give similar testimony. My own father-well, as I have already said, my own father would have drowned himself had it not been for my mother's prayers and faith. Probably thousands of the tortured souls who are now screaming in our insane asylums could have been saved if they had only turned to a higher power for help instead of trying to fight life's battles alone.

When we are harassed and reach the limit of our own strength, many of us then turn in desperation to God-"There are no atheists in foxholes." But why wait till we are desperate? Why not renew our strength every day? Why wait even until Sunday? For years I have had the habit of dropping into empty churches on weekday afternoons. When I feel that I am too rushed and hurried to spare a few minutes to think about spiritual things, I say to myself: "Wait a minute, Dale Carnegie, wait a minute. Why all the feverish hurry and rush, little man? You need to pause and acquire a little perspective." At such times, I frequently drop into the first church that I find open. Although I am a Protestant, I frequently, on weekday afternoons, drop into St. Patrick's Cathedral on Fifth Avenue, and remind myself that I'll be dead in another thirty years, but that the great spiritual truths that all churches teach are eternal. I close my eyes and pray. I find that doing this calms my nerves, rests my body, clarifies my per spective, and helps me revalue my values. May I recommend this practice to you?

During the past six years that I have been writing this book I have collected hundreds of examples and concrete cases of how men and women conquered fear and worry by prayer. I have in my filing cabinet folders bulging with case histories. Let's take as a typical example the story of a discouraged and disheartened book salesman, John R. Anthony. Mr. Anthony is now an attorney in Houston, Texas, with offices in the Humble Building. Here is his story as he told it to me.

"Twenty-two years ago I closed my private law office to become state representative of an American law-book company. My specialty was selling a set of law-books to lawyers-a set of books that were almost indispensable.

"I was ably and thoroughly trained for the job. I knew all the direct sales talks, and the convincing answers to all possible objections. Before calling on a prospect, I familiarised myself with his rating as an attorney, the nature of his practice, his politics and hobbies. During my interview, I used that information with ample skill. Yet, something was wrong. I just couldn't get orders!

"I grew discouraged. As the days and weeks passed, I doubled and redoubled ray efforts, but was still unable to close enough sales to pay my expenses. A sense of fear and dread grew within me. I became afraid to call on people. Before I could enter a prospect's office, that feeling of dread flared up so strong that I would pace up and down the hallway outside the door-or go out of the building and circle the block. Then, after losing much valuable time and feigning enough courage by sheer will power to crash the office door, I feebly turned the doorknob with trembling hand-half hoping my prospect would not be in!

"My sales manager threatened to stop my advances if I didn't send in more orders. My wife at home pleaded with me for money to pay the grocery bill for herself and our three children. Worry seized me. Day by day I grew more desperate. I didn't know what to do. As I have already said, I had closed my private law office at home and given up my clients. Now I was broke. I didn't have the money to pay even my hotel bill. Neither did I have the money to buy a ticket back home; nor did I have the courage to return home a beaten man, even if I had had the ticket. Finally, at the miserable end of another bad day, I trudged back to my hotel room-for the last time, I thought. So far as I was concerned, I was thoroughly beaten.

Heartbroken, depressed, I didn't know which way to turn. I hardly cared whether I lived or died. I was sorry I had ever been born. I had nothing but a glass of hot milk that night for dinner. Even that was more than I could afford. I understood that night why desperate men raise a hotel window and jump. I might have done it myself if I had had the courage. I began wondering what was the purpose of life. I didn't know. I couldn't figure it out.

"Since there was no one else to turn to, I turned to God. I began to pray. I implored the Almighty to give me light and understanding and guidance through the dark, dense wilderness of despair that had closed in about me. I asked God to help me get orders for my books and to give me money to feed my wife and children. After that prayer, I opened my eyes and saw a Gideon Bible that lay on the dresser in that lonely hotel room. I opened it and read those beautiful, immortal promises of Jesus that must have inspired countless generations of lonely, worried, and beaten men throughout the ages-a talk that Jesus gave to His disciples about how to keep from worrying:

Take no thought for your life, what ye shall eat, or what ye shall drink; not yet for your body, what ye shall put on. Is not the life more than meat, and the body than raiment? Behold the fowls of the air: for they sow not, neither do they reap, nor gather into barns; yet your heavenly Father feedeth them. Are ye not much better than they? ... But seek ye first the kingdom of God, and his righteousness; and all these things shall be added unto you.

"As I prayed and as I read those words, a miracle happened: my nervous tension fell away. My anxieties, fears, and worries were transformed into heart-warming courage and hope and triumphant faith.

"I was happy, even though I didn't have enough money to pay my hotel bill. I went to bed and slept soundly-free from care-as I had not done for many years.

"Next morning, I could hardly hold myself back until the offices of my prospects were open. I approached the office door of my first prospect that beautiful, cold, rainy day with a bold and positive stride. I turned the doorknob with a firm and steady grip. As I entered, I made a beeline for my man, energetically, chin up, and with appropriate dignity, all smiles, and saying: 'Good morning, Mr. Smith! I'm John R. Anthony of the All-American Lawbook Company!'

" 'Oh, yes, yes,' he replied, smiling, too, as he rose from his chair with outstretched hand. 'I'm glad to see you. Have a seat!'

"I made more sales that day than I had made in weeks. That evening I proudly returned to my hotel like a conquering hero! I felt like a new man. And I was a new man, because I had a new and victorious mental attitude. No dinner of hot milk that night. No, sir! I had a steak with all the fixin's. From that day on, my sales zoomed.

"I was born anew that desperate night twenty-one years ago in a little hotel in Amarillo, Texas. My outward situation the next day was the same as it had been through my weeks of failure, but a tremendous thing had happened inside me. I had suddenly become aware of my relationship with God. A mere man alone can easily be defeated, but a man alive with the power of God within him is invincible. I know. I saw it work in my own life.

" 'Ask, and it shall be given you; seek, and ye shall find; knock, and it shall be opened unto you.' "

When Mrs. L. G. Beaird, of 1421 8th Street, Highland, Illinois, was faced with stark tragedy, she discovered that she could find peace and tranquility by kneeling down and saying: "O Lord, Thy will, not mine, be done."

"One evening our telephone rang," she writes in a letter that I have before me now. "It rang fourteen times before I had the courage to pick up the receiver. I knew it must be the hospital, and I was terrified. I feared that our little boy was dying. He had meningitis. He had already been given penicillin, but it made his temperature fluctuate, and the doctor feared that the disease had travelled to his brain and might cause the development of a brain tumour-and death. The phone call was just what I feared. The hospital was calling; the doctor wanted us to come immediately.

"Maybe you can picture the anguish my husband and I went through, sitting in the waiting-room. Everyone else had his baby, but we sat there with empty arms, wondering if we would ever hold our little fellow again. When we were finally called into the doctor's private office, the expression on his face filled our heart with terror. His words brought even more terror. He told us that there was only one chance in four that our baby would live. He said that if we knew another doctor, to please call him in on the case.

"On the way home my husband broke down and, doubling up his fist, hit the steering wheel, saying: 'Berts, I can't give that little guy up.' Have you ever seen a man cry? It isn't a pleasant experience. We stopped the car and, after talking things over, decided to stop in church and pray that if it was God's will to take our baby, we would resign our will to His. I sank in the pew and said with tears rolling down my cheeks: 'Not my will but Thine be done.'

"The moment I uttered those words, I felt better. A sense of peace that I hadn't felt for a long time came over me. All the way home, I kept repeating: 'O God, Thy will, not mine, be done.'

"I slept soundly that night for the first time in a week. The doctor called a few days later and said that Bobby had passed the crisis. I thank God for the strong and healthy four-year-old boy we have today."

I know men who regard religion as something for women and children and preachers. They pride themselves on being "he-men" who can fight their battles alone.

How surprised they might be to learn that some of the most famous "he-men" in the world pray every day. For example, "he-man" Jack Dempsey told me that he never goes to bed without saying his prayers. He told me that he never eats a meal without first thanking God for it. He told me that he prayed every day when he was training for a bout, and that when he was fighting, he always prayed just before the bell sounded for each round. "Praying," he said, "helped me fight with courage and confidence."

"He-man" Connie Mack told me that he couldn't go to sleep without saying his prayers.

"He-man" Eddie Rickenbacker told me that he believed his life had been saved by prayer. He prays every day.

"He-man" Edward R. Stettinius, former high official of General Motors and United States Steel, and former Secretary of State, told me that he prayed for wisdom and guidance every morning and night.

"He-man" J. Pierpont Morgan, the greatest financier of his age, often went alone to Trinity Church, at the head of Wall Street, on Saturday afternoons and knelt in prayer.

When "he-man" Eisenhower flew to England to take supreme command of the British and American forces, he took only one book on the plane with him-the Bible.

"He-man" General Mark Clark told me that he read his Bible every day during the war and knelt down in prayer. So did Chiang Kai-shek, and General Montgomery-"Monty of El Alamein". So did Lord Nelson at Trafalgar. So did General Washington, Robert E. Lee, Stonewall Jackson, and scores of other great military leaders.

These "he-men" discovered the truth of William James's statement: "We and God have business with each other; and in opening ourselves to His influence, our deepest destiny is fulfilled."

A lot of "he-men" are discovering that. Seventy-two million Americans are church members now-an all-time record. As I said before, even the scientists are turning to religion. Take, for example, Dr. Alexis Carrel, who wrote *Man, the Unknown* and won the greatest honour that can be bestowed upon any scientist, the Nobel prize. Dr. Carrel said in a Reader's Digest article:

"Prayer is the most powerful form of energy one can generate. It is a force as real as terrestrial gravity. As a physician, I have seen men, after all other therapy had failed, lifted out of disease and melancholy by the serene effort of prayer. ... Prayer like radium is a source of luminous, self-generating energy. ... In prayer, human beings seek to augment their finite energy by addressing themselves to the Infinite source of all energy. When we pray, we link ourselves with the inexhaustible motive power that spins the universe. We pray that a part of this power be apportioned to our needs.

"Even in asking, our human deficiencies are filled and we arise strengthened and repaired. ... Whenever we address God in fervent prayer, we change both soul and body for the better. It could not happen that any man or woman could pray for a single moment without some good result."

Admiral Byrd knows what it means to "link ourselves with the inexhaustible motive power that spins the universe". His ability to do that pulled him through the most trying ordeal of his life. He tells the story in his book *Alone*. (*) In 1934, he spent five months in a hut buried beneath the icecap of Ross Barrier deep in the Antarctic. He was the only living creature south of latitude seventy-eight. Blizzards roared above his shack; the cold plunged down to eighty-two degrees below zero; he was completely surrounded by unending night. And then he found, to his horror, he was being slowly poisoned by carbon monoxide that escaped from his stove! What could he do? The nearest help was 123 miles away, and could not possibly reach him for several months. He tried to fix his stove and ventilating system, but the fumes still escaped. They often knocked him out cold. He lay on the floor completely unconscious. He couldn't eat; he couldn't sleep; he became so feeble that he could hardly leave his bunk. He frequently feared he wouldn't live until morning. He was convinced he would die in that cabin, and his body would be hidden by perpetual snows.

---

* Putnam & Co. Ltd.

What saved his life? One day, in the depths of his despair, he reached for his diary and tried to set down his philosophy of life. "The human race," he wrote, "is not alone in the universe." He thought of the stars overhead, of the orderly swing of the constellations and planets; of how the everlasting sun would, in its time, return to lighten even the wastes of the South Polar regions. And then he wrote in his diary: "I am not alone."

This realisation that he was not alone-not even in a hole in the ice at the end of the earth-was what saved Richard Byrd. "I know it pulled me through," he says. And he goes on to add: "Few men in their lifetime come anywhere near exhausting the resources dwelling within them. There are deep wells of strength that are never used." Richard Byrd learned to tap those wells of strength and use those resources-by turning to God.

Glenn A. Arnold learned amidst the cornfields of Illinois the same lesson that Admiral Byrd learned in the polar icecap. Mr. Arnold, an insurance broker in the Bacon Building, Chillicothe, Illinois, opened his speech on conquering worry like this: "Eight years ago, I turned the key in the lock of my front door for what I believed was the last time in my life. I then climbed in my car and started down for the river. I was a failure," he said. "One month before, my entire little world had come crashing down on my head. My electrical-appliance business had gone on the rocks. In my home my mother lay at the point of death. My wife was carrying our second child. Doctors' bills were mounting. We had mortgaged everything we had to start the business-our car and our furniture. I had even taken out a loan on my insurance policies. Now everything was gone. I couldn't take it any longer. So I climbed into my car and started for the river-determined to end the sorry mess.

"I drove a few miles out in the country, pulled off the road, and got out and sat on the ground and wept like a child. Then I really started to think-instead of going around in frightening circles of worry, I tried to think constructively. How bad was my situation? Couldn't it be worse? Was it really hopeless? What could I do to make it better?

"I decided then and there to take the whole problem to the Lord and ask Him to handle it. I prayed. I prayed hard. I prayed as though my very life depended on it-which, in fact, it did. Then a strange thing happened. As soon as I turned all my problems over to a power greater than myself, I immediately felt a peace of mind that I hadn't known in months. I must have sat there for half an hour, weeping and praying. Then I went home and slept like a child.

"The next morning, I arose with confidence. I no longer had anything to fear, for I was depending on God for guidance. That morning I walked into a local department store with my head high; and I spoke with confidence as I applied for a job as salesman in the electrical-appliance department. I knew I would get a job. And I did. I made good at it until the whole appliance business collapsed due to the war. Then I began selling life insurance-still under the management of my Great Guide. That was only five years ago. Now, all my bills are paid; I have a fine family of three bright children; own my own home; have a new car, and own twenty-five thousand dollars in life insurance.

"As I look back, I am glad now that I lost everything and became so depressed that I started for the river-because that tragedy taught me to rely on God; and I now have a peace and confidence that I never dreamed were possible."

Why does religious faith bring us such peace and calm and fortitude? I'll let William James answer that. He says: "The turbulent billows of the fretful surface leave the deep parts of the ocean undisturbed; and to him who has a hold on vaster and more permanent realities, the hourly vicissitudes of his personal destiny seem relatively insignificant things. The really religious person is accordingly unshakable and full of equanimity, and calmly ready for any duty that the day may bring forth."

If we are worried and anxious-why not try God ? Why not, as Immanuel Kant said: "accept a belief in God because we need such a belief"? Why not link ourselves now "with the inexhaustible motive power that spins the universe"?

Even if you are not a religious person by nature or training-even if you are an out-and-out sceptic prayer can help you much more than you believe, for it is a practical thing. What do I mean, practical? I mean that prayer fulfills these three very basic psychological needs which all people share, whether they believe in God or not:

1. Prayer helps us to put into words exactly what is troubling us. We saw in Chapter 4 that it is almost impossible to deal with a problem while it remains vague and nebulous. Praying, in a way, is very much like writing our problem down on paper. If we ask help for a problem-even from God-we must put it into words.

2. Prayer gives us a sense of sharing our burdens, of not being alone. Few of us are so strong that we can bear our heaviest burdens, our most agonising troubles, all by ourselves. Sometimes our worries are of so intimate a nature that we cannot discuss them even with our closest relatives or friends. Then prayer is the answer. Any psychiatrist will tell us that when we are pent-up and tense, and in an agony of spirit, it is therapeutically good to tell someone our troubles. When we can't tell anyone else-we can always tell God.

3. Prayer puts into force an active principle of doing. It's a first step toward action. I doubt if anyone can pray for some fulfillment, day after day, without benefiting from it-in other words, without taking some steps to bring it to pass. A world-famous scientist said: "Prayer is the most powerful form of energy one can generate." So why not make use of it? Call it God or Allah or Spirit-why quarrel with definitions as long as the mysterious powers of nature take us in hand?

Why not close this book right now, go to your bedroom, shut the door, kneel down, and unburden your heart? If you have lost your religion, beseech Almighty God to renew your faith. Say: "O God, I can no longer fight my battles alone. I need Your help, Your love. Forgive me for all my mistakes. Cleanse my heart of all evil. Show me the way to peace and quiet and health, and fill me with love even for my enemies."

If you don't know how to pray, repeat this beautiful and inspiring prayer written by St. Francis seven hundred years ago:

Lord, make me an instrument of Thy Peace. Where there is hatred, let me sow love; where there is injury, pardon; where there is doubt, faith; where there is despair, hope; where there is darkness, light; where there is sadness, joy.

O Divine Master, grant that I may not so much seek to be consoled as to console; to be understood, as to understand; to be loved, as to love; for it is in giving that we receive, it is in pardoning, that we are pardoned and it is in dying that we are born to Eternal Life.

✦ ✦ ✦

PART SIX

How To Keep From Worrying About Criticism

Chapter 20

Remember That No One Ever Kicks A Dead Dog

An event occurred in 1929 that created a national sensation in educational circles. Learned men from all over America rushed to Chicago to witness the affair. A few years earlier, a young man by the name of Robert Hutchins had worked his way through Yale, acting as a waiter, a lumberjack, a tutor, and a clothes-line salesman. Now, only eight years later, he was being inaugurated as president of the fourth richest university in America, the University of Chicago. His age? Thirty. Incredible! The older educators shook their heads. Criticism came roaring down upon the "boy wonder" like a rockslide. He was this and he was that-too young, inexperienced-his educational ideas were cockeyed. Even the newspapers joined in the attack.

The day he was inaugurated, a friend said to the father of Robert Maynard Hutchins: "I was shocked this morning to read that newspaper editorial denouncing your son."

"Yes," the elder Hutchins replied, "it was severe, but remember that no one ever kicks a dead dog."

Yes, and the more important a dog is, the more satisfaction people get in kicking him. The Prince of Wales who later became Edward VIII (now Duke of Windsor) had that forcibly brought home to him. He was attending Dartmouth College in Devonshire at the time-a college that corresponds to the Naval Academy at Annapolis. The Prince was about fourteen. One day one of the naval officers found him crying, and asked him what was wrong. He refused to tell at first, but finally admitted the truth: he was being kicked by the naval cadets. The commodore of the college summoned the boys and explained to them that the Prince had not complained, but he wanted to find out why the Prince had been singled out for this rough treatment.

After much hemming and hawing and toe scraping, the cadets finally confessed that when they themselves became commanders and captains in the King's Navy, they wanted to be able to say that they had kicked the King!

So when you are kicked and criticised, remember that it is often done because it gives the kicker a feeling of importance. It often means that you are accomplishing something and are worthy of attention. Many people get a sense of savage satisfaction out of denouncing those who are better educated than they are or more successful. For example, while I was writing this chapter, I received a letter from a woman denouncing General William Booth, founder of the Salvation Army. I had given a laudatory broadcast about General Booth; so this woman wrote me, saying that General Booth had stolen eight million dollars of the money he had collected to help poor people. The charge, of course, was absurd. But this woman wasn't looking for truth. She was seeking the mean-spirited gratification that she got from tearing down someone far above her. I threw her bitter letter into the wastebasket, and thanked Almighty God that I wasn't married to her. Her letter didn't tell me anything at all about General Booth, but it did tell me a lot about her. Schopenhauer had said it years ago: "Vulgar people take huge delight in the faults and follies of great men."

One hardly thinks of the president of Yale as a vulgar man; yet a former president of Yale, Timothy Dwight, apparently took huge delight in denouncing a man who was running for President of the United States. The president of Yale warned that if this man were elected President, "we may see our wives and daughters the victims of legal prostitution, soberly dishonoured, speciously polluted; the outcasts of delicacy and virtue, the loathing of God and man."

Sounds almost like a denunciation of Hitler, doesn't it? But it wasn't. It was a denunciation of Thomas Jefferson. Which Thomas Jefferson? Surely not the immortal Thomas Jefferson, the author of the Declaration of Independence, the patron saint of democracy? Yea, verily, that was the man.

What American do you suppose was denounced as a "hypocrite", "an impostor", and as "little better than a murderer"?

A newspaper cartoon depicted him on a guillotine, the big knife ready to cut off his head. Crowds jeered at him and hissed him as he rode through the street. Who was he? George Washington.

But that occurred a long time ago. Maybe human nature has improved since then. Let's see. Let's take the case of Admiral Peary-the explorer who startled and thrilled the world by reaching the North Pole with dog sleds on April 6, 1909-a goal that brave men for centuries had suffered and died to attain. Peary himself almost died from cold and starvation; and eight of his toes were frozen so hard they had to be cut off. He was so overwhelmed with disasters that he feared he would go insane. His superior naval officers in Washington were burned up because Peary was getting so much publicity and acclaim. So they accused him of collecting money for scientific expeditions and then "lying around and loafing in the Arctic." And they probably believed it, because it is almost impossible not to believe what you want to believe. Their determination to humiliate and block Peary was so violent that only a direct order from President McKinley enabled Peary to continued his career in the Arctic.

Would Peary have been denounced if he had had a desk job in the Navy Department in Washington? No. He wouldn't have been important enough then to have aroused jealousy.

General Grant had an even worse experience than Admiral Peary. In 1862, General Grant won the first great decisive victory that the North had enjoyed-a victory that was achieved in one afternoon, a victory that made Grant a national idol overnight-a victory that had tremendous repercussions even in far-off Europe-a victory that set church bells ringing and bonfires blazing from Maine to the banks of the Mississippi. Yet within six weeks after achieving that great victory, Grant -hero of the North-was arrested and his army was taken from him. He wept with humiliation and despair.

Why was General U.S. Grant arrested at the flood tide of his victory? Largely because he had aroused the jealousy and envy of his arrogant superiors.

If we are tempted to be worried about unjust criticism here is Rule 1:

6.1

RULE 1

Remember that unjust criticism is often a disguised compliment. Remember that no one ever kicks a dead dog.

✦ ✦ ✦

Chapter 21

Do This-and Criticism Can't Hurt You

I once interviewed Major-General Smedley Butler-old "Gimlet-Eye". Old "Hell-Devil" Butler! Remember him? The most colourful, swashbuckling general who ever commanded the United States Marines.

He told me that when he was young, he was desperately eager to be popular, wanted to make a good impression on everyone. In those days the slightest criticism smarted and stung. But he confessed that thirty years in the Marines had toughened his hide. "I have been berated and insulted," he said, "and denounced as a yellow dog, a snake, and a skunk. I have been cursed by the experts. I have been called every possible combination of unprintable cuss words in the English language. Bother me? Huh! When I hear someone cussing me now, I never turn my head to see who is talking."

Maybe old "Gimlet-Eye" Butler was too indifferent to criticism; but one thing is sure: most of us take the little jibes and javelins that are hurled at us far too seriously. I remember the time, years ago, when a reporter from the New York Sun attended a demonstration meeting of my adult-education classes and lampooned me and my work. Was I burned up? I took it as a personal insult. I telephoned Gill Hodges, the Chairman of the Executive Committee of the Sun, and practically demanded that he print an article stating the facts-instead of ridicule. I was determined to make the punishment fit the crime.

I am ashamed now of the way I acted. I realise now that half the people who bought the paper never saw that article. Half of those who read it regarded it as a source of innocent merriment. Half of those who gloated over it forgot all about it in a few weeks.

I realise now that people are not thinking about you and me or caring what is said about us. They are thinking about themselves-before breakfast, after breakfast, and right on until ten minutes past midnight. They would be a thousand times more concerned about a slight headache of their own than they would about the news of your death or mine.

Even if you and I are lied about, ridiculed, double-crossed, knifed in the back, and sold down the river by one out of every six of our most intimate friends-let's not indulge in an orgy of self-pity. Instead, let's remind ourselves that that's precisely what happened to Jesus. One of His twelve most intimate friends turned traitor for a bribe that would amount, in our modern money, to about nineteen dollars. Another one of His twelve most intimate friends openly deserted Jesus the moment He got into trouble, and declared three times that he didn't even know Jesus-and he swore as he said it. One out of six! That is what happened to Jesus. Why should you and I expect a better score?

I discovered years ago that although I couldn't keep people from criticising me unjustly, I could do something infinitely more important: I could determine whether I would let the unjust condemnation disturb me.

Let's be clear about this: I am not advocating ignoring all criticism. Far from it. I am talking about ignoring only unjust criticism. I once asked Eleanor Roosevelt how she handled unjust criticism-and Allah knows she's had a lot of it. She probably has more ardent friends and more violent enemies than any other woman who ever lived in the White House.

She told me that as a young girl she was almost morbidly shy, afraid of what people might say. She was so afraid of criticism that one day she asked her aunt, Theodore Roosevelt's sister for advice. She said: "Auntie Bye, I want to do so-and-so. But I'm afraid of being criticised."

Teddy Roosevelt's sister looked her in the eye and said: "Never be bothered by what people say, as long as you know in your heart you are right." Eleanor Roosevelt told me that that bit of advice proved to be her Rock of Gibraltar years later, when she was in the White House. She told me that the only way we can avoid all criticism is to be like a Dresden-china figure and stay on a shelf. "Do what you feel in your heart to be right-for you'll be criticised, anyway. You'll be damned if you do, and damned if you don't." That is her advice.

When the late Matthew C. Brush, was president of the American International Corporation at 40 Wall Street, I asked him if he was ever sensitive to criticism; and he replied: "Yes, I was very sensitive to it in my early days. I was eager then to have all the employees in the organisation think I was perfect. If they didn't, it worried me. I would try to please first one person who had been sounding off against me; but the very thing I did to patch it up with him would make someone else mad. Then when I tried to fix it up with this person, I would stir up a couple of other bumble-bees. I finally discovered that the more I tried to pacify and to smooth over injured feelings in order to escape personal criticism, the more certain I was to increase my enemies. So finally I said to myself: 'If you get your head above the crowd, you're going to be criticised. So get used to the idea.' That helped me tremendously. From that time on I made it a rule to do the very best I could and then put up my old umbrella and let the rain of criticism drain off me instead of running down my neck."

Deems Taylor went a bit further: he let the rain of criticism run down his neck and had a good laugh over it-in public. When he was giving his comments during the intermission of the Sunday afternoon radio concerts of the New York Philharmonic-Symphony Orchestra, one woman wrote him a letter calling him "a liar, a traitor, a snake and a moron".

On the following week's broadcast, Mr. Taylor read this letter over the radio to millions of listeners. In his book, *Of Men & Music*, he tells us that a few days later he received another letter from the same lady, "expressing her unaltered opinion that I was still a liar, a traitor, a snake and a moron. I have a suspicion," adds Mr. Taylor, "that she didn't care for that talk." We can't keep from admiring a man who takes criticism like that. We admire his serenity, his unshaken poise, and his sense of humour.

When Charles Schwab was addressing the student body at Princeton, he confessed that one of the most important lessons he had ever learned was taught to him by an old German who worked in Schwab's steel mill. The old German got involved in a hot wartime argument with the other steelworkers, and they tossed him into the river. "When he came into my office," Mr. Schwab said, "covered with mud and water, I asked him what he had said to the men who had thrown him into the river, and he replied: 'I just laughed.'"

Mr. Schwab declared that he had adopted that old German's words as his motto: "Just laugh." That motto is especially good when you are the victim of unjust criticism. You can answer the man who answers you back, but what can you say to the man who "just laughs"?

Lincoln might have broken under the strain of the Civil War if he hadn't learned the folly of trying to answer all his savage critics. He finally said: "If I were to try to read, much less to answer, all the attacks made on me, this shop might as well be closed for any other business. I do the very best I know how- the very best I can; and I mean to keep on doing so until the end. If the end brings me out all right, then what is said against me won't matter. If the end brings me out wrong, then ten angels swearing I was right would make no difference."

When you and I are unjustly criticised, let's remember Rule 2:

6.2

RULE 2

Do the very best you can: and then put up your old umbrella and keep the rain of criticism from running down the back of your neck.

✣ ✣ ✣

Chapter 22

Fool Things I Have Done

I have a folder in my private filing cabinet marked "FTD"-short for "Fool Things I Have Done". I put in that folder written records of the fools things I have been guilty of. I sometimes dictate these memos to my secretary, but sometimes they are so personal, so stupid, that I am ashamed to dictate them, so I write them out in longhand.

I can still recall some of the criticisms of Dale Carnegie that I put in my "FTD" folders fifteen years ago. If I had been utterly honest with myself, I would now have a filing cabinet bursting out at the seams with these "FTD" memos. I can truthfully repeat what King Saul said more than twenty centuries ago: "I have played the fool and have erred exceedingly."

When I get out my "FTD" folders and re-read the criticisms I have written of myself, they help me deal with the toughest problem I shall ever face: the management of Dale Carnegie.

I used to blame my troubles on other people; but as I have grown older-and wiser, I hope-I have realised that I myself, in the last analysis, am to blame for almost all my misfortunes. Lots of people have discovered that, as they grow older. "No one but myself," said Napoleon at St. Helena, "no one but myself can be blamed for my fall. I have been my own greatest enemy-the cause of my own disastrous fate."

Let me tell you about a man I know who was an artist when it came to self-appraisal and selfmanagement. His name was H. P. Howell. When the news of his sudden death in the drugstore of the Hotel Ambassador in New York was flashed across the nation on July 31, 1944, Wall Street was shocked, for he was a leader in American finance-chairman of the board of the Commercial National Bank and Trust Company, 56 Wall Street, and a director of several large corporations. He grew up with little formal education, started out in life clerking in a country store, and later became credit manager for U.S. Steel- and was on his way to position and power.

"For years I have kept an engagement book showing all the appointments I have during the day," Mr. Howell told me when I asked him to explain the reasons for his success. "My family never makes any plans for me on Saturday night, for the family knows that I devote a part of each Saturday evening to self-examination and a review and appraisal of my work during the week. After dinner I go off by myself, open my engagement book, and think over all the interviews, discussions and meetings that have taken place since Monday morning. I ask myself: 'What mistakes did I make that time?' 'What did I do that was right-and in what way could I have improved my performance?' 'What lessons can I learn from that experience?' I sometimes find that this weekly review makes me very unhappy. Sometimes I am astonished by my own blunders. Of course, as the years have gone by, these blunders have become less frequent. This system of self-analysis, continued year after year, has done more for me than any other one thing I have ever attempted."

Maybe H.P. Howell borrowed his idea from Ben Franklin. Only Franklin didn't wait until Saturday night. He gave himself a severe going-over every night. He discovered that he had thirteen serious faults. Here are three of them: wasting time, stewing around over trifles, arguing and contradicting people. Wise old Ben Franklin realised that, unless he eliminated these handicaps, he wasn't going to get very far. So he battled with one of his shortcomings every day for a week, and kept a record of who had won each day's slugging match. The next day, he would pick out another bad habit, put on the gloves, and when the bell rang he would come out of his corner fighting. Franklin kept up this battle with his faults every week for more than two years.

No wonder he became one of the best-loved and most influential men America ever produced!

Elbert Hubbard said: "Every man is a damn fool for at least five minutes every day. Wisdom consists in not exceeding that limit."

The small man flies into a rage over the slightest criticism, but the wise man is eager to learn from those who have censured him and reproved him and "disputed the passage with him". Walt Whitman put it this way: "Have you learned lessons only of those who admired you, and were tender with you, and stood aside for you? Have you not learned great lessons from those who rejected you, and braced themselves against you, or disputed the passage with you?"

Instead of waiting for our enemies to criticise us or our work, let's beat them to it. Let's be our own most severe critic. Let's find and remedy all our weaknesses before our enemies get a chance to say a word. That is what Charles Darwin did. In fact, he spent fifteen years criticising-well, the story goes like this: When Darwin completed the manuscript of his immortal book, The Origin of Species, he realised that the publication of his revolutionary concept of creation would rock the intellectual and religious worlds. So he became his own critic and spent another fifteen years, checking his data, challenging his reasoning, criticising his conclusions.

Suppose someone denounced you as "a damn fool"-what would you do? Get angry? Indignant? Here is what Lincoln did: Edward M. Stanton, Lincoln's Secretary of War, once called Lincoln "a damn fool". Stanton was indignant because Lincoln had been meddling in his affairs. In order to please a selfish politician, Lincoln had signed an order transferring certain regiments. Stanton not only refused to carry out Lincoln's orders but swore that Lincoln was a damn fool forever signing such orders. What happened? When Lincoln was told what Stanton had said, Lincoln calmly replied: "If Stanton said I was a damned fool, then I must be, for he is nearly always right. I'll just step over and see for myself."

Lincoln did go to see Stanton. Stanton convinced him that the order was wrong, and Lincoln withdrew it. Lincoln welcomed criticism when he knew it was sincere, founded on knowledge, and given in a spirit of helpfulness.

You and I ought to welcome that kind of criticism, too, for we can't even hope to be right more than three times out of four. At least, that was all Theodore Roosevelt said he could hope for, when he was in the White House. Einstein, the most profound thinker now living, confesses that his conclusions are wrong ninety-nine per cent of the time!

"The opinions of our enemies," said La Rochefoucauld, "come nearer to the truth about us than do our own opinions."

I know that statement may be true many times; yet when anyone starts to criticise me, if I do not watch myself, I instantly and automatically leap to the defensive-even before I have the slightest idea what my critic is going to say. I am disgusted with myself every time I do it. We all tend to resent criticism and lap up praise, regardless of whether either the criticism or the praise be justified. We are not creatures of logic. We are creatures of emotions. Our logic is like a canoe tossed about on a deep, dark, stormy sea of emotion. Most of us have a pretty good opinion of ourselves as we are now. But in forty years from now, we may look back and laugh at the persons we are today.

William Allen White - the most celebrated small-town newspaper editor in history - looked back and described the young man he had been fifty years earlier as "swell-headed ... a fool with a lot of nerve ... a supercilious young Pharisee ... a complacent reactionary." Twenty years from now maybe you and I may be using similar adjectives to describe the persons we are today. We may. ... who knows?

In previous chapters, I have talked about what to do when you are unjustly criticised. But here is another idea: when your anger is rising because you feel you have been unjustly condemned, why not stop and say: "Just a minute. ... I am far from perfect. If Einstein admits he is wrong ninety-nine per cent of the time, maybe I am wrong at least eighty per cent of the time. Maybe I deserve this criticism. If I do, I ought to be thankful for it, and try to profit by it."

Charles Luckman, president of the Pepsodent Company, spends a millions dollars a year putting Bob Hope on the air. He doesn't look at the letters praising the programme, but he insists on seeing the critical letters. He knows he may learn something from them.

The Ford Company is so eager to find out what is wrong with its management and operations that it recently polled the employees and invited them to criticise the company.

I know a former soap salesman who used even to ask for criticism. When he first started out selling soap for Colgate, orders came slowly. He worried about losing his job. Since he knew there was nothing wrong with the soap or the price, he figured that the trouble must be himself. When he failed to make a sale, he would often walk around the block trying to figure out what was wrong. Had he been too vague? Did he lack enthusiasm? Sometimes he would go back to the merchant and say: "I haven't come back here to try to sell you any soap. I have come back to get your advice and your criticism. Won't you please tell me what I did that was wrong when I tried to sell you soap a few minutes ago? You are far more experienced and successful than I am. Please give me your criticism. Be frank. Don't pull your punches."

This attitude won him a lot of friends and priceless advice.

What do you suppose happened to him? Today, he is president of the Colgate-Palmolive-Peet Soap Company-the world's largest makers of soap. His name is E. H. Little. Last year, only fourteen people in America had a larger income than he had: $240,141.

It takes a big man to do what H. P. Howell, Ben Franklin, and E. H. Little did. And now, while nobody is looking, why not peep into the mirror and ask yourself whether you belong in that kind of company?

To keep from worrying about criticism, here is Rule 3:

6.3

RULE 3

Let's keep a record of the fool things we have done and criticise ourselves. Since we can't hope to be perfect, let's do what E.H. Little did: let's ask for unbiased, helpful, constructive criticism.

Part Six In A Nutshell - How To Keep From Worrying About Criticism

RULE 1 Unjust criticism is often a disguised compliment. It often means that you have aroused jealousy and envy. Remember that no one ever kicks a dead dog.

RULE 2 Do the very best you can; and then put up your old umbrella and keep the rain of criticism from running down the back of your neck.

RULE 3 Let's keep a record of the fool things we have done and criticise ourselves. Since we can't hope to be perfect, let's do what E. H. Little did: let's ask for unbiased, helpful, constructive criticism.

✦ ✦ ✦

PART SEVEN

Six Ways To Prevent Fatigue And Worry And Keep Your Energy And Spirits High

Chapter 23

How To Add One Hour A Day To Your Waking Life

Why am I writing a chapter on preventing fatigue in a book on preventing worry? That is simple: because fatigue often produces worry, or, at least, it makes you susceptible to worry. Any medical student will tell you that fatigue lowers physical resistance to the common cold and hundreds of other diseases and any psychiatrist will tell you that fatigue also lowers your resistance to the emotions of fear and worry. So preventing fatigue tends to prevent worry.

Did I say "tends to prevent worry"? That is putting it mildly. Dr. Edmund Jacobson goes much further. Dr. Jacobson has written two books on relaxation: *Progressive Relaxation and You Must Relax*, and as director of the University of Chicago Laboratory for Clinical Physiology, he has spent years conducting investigations in using relaxation as a method in medical practice. He declares that any nervous or emotional state "fails to exist in the presence of complete relaxation". That is another way of saying: You cannot continue to worry if you relax.

So, to prevent fatigue and worry, the first rule is: Rest often. Rest before you get tired.

Why is that so important? Because fatigue accumulates with astonishing rapidity. The United States Army has discovered by repeated tests that even young men-men toughened by years of Army training can march better, and hold up longer, if they throw down their packs and rest ten minutes out of every hour. So the Army forces them to do just that. Your heart is just as smart as the U.S. Army. Your heart pumps enough blood through your body every day to fill a railway tank car. It exerts enough energy every twenty-four hours to shovel twenty tons of coal on to a platform three feet high. It does this incredible amount of work for fifty, seventy, or maybe ninety years. How can it stand it? Dr. Walter B. Cannon, of the Harvard Medical School, explains it. He says: "Most people have the idea that the heart is working all the time. As a matter of fact, there is a definite rest period after each contraction. When beating at a moderate rate of seventy pulses per minute, the heart is actually working only nine hours out of the twenty-four. In the aggregate its rest periods total a full fifteen hours per day."

During World War II, Winston Churchill, in his late sixties and early seventies, was able to work sixteen hours a day, year after year, directing the war efforts of the British Empire. A phenomenal record. His secret? He worked in bed each morning until eleven o'clock, reading papers, dictating orders, making telephone calls, and holding important conferences. After lunch he went to bed once more and slept for an hour. In the evening he went to bed once more and slept for two hours before having dinner at eight. He didn't cure fatigue. He didn't have to cure it. He prevented it. Because he rested frequently, he was able to work on, fresh and fit, until long past midnight.

The original John D. Rockefeller made two extraordinary records. He accumulated the greatest fortune the world had ever seen up to that time and he also lived to be ninety-eight. How did he do it? The chief reason, of course, was because he had inherited a tendency to live long. Another reason was his habit of taking a half-hour nap in his office every noon. He would lie down on his office couch-and not even the President of the United States could get John D. on the phone while he was having his snooze!

In his excellent book. *Why Be Tired*, Daniel W. Josselyn observes: "Rest is not a matter of doing absolutely nothing. Rest is repair." There is so much repair power in a short period of rest that even a five-minute nap will help to forestall fatigue! Connie Mack, the grand old man of baseball, told me that if he doesn't take an afternoon nap before a game, he is all tuckered out at around the fifth inning. But if he does go to sleep, if for only five minutes, he can last throughout an entire double-header without feeling tired.

When I asked Eleanor Roosevelt how she was able to carry such an exhausting schedule during the twelve years she was in the White House, she said that before meeting a crowd or making a speech, she would often sit in a chair or davenport, close her eyes, and relax for twenty minutes.

I recently interviewed Gene Autry in his dressing-room at Madison Square Garden, where he was the star attraction at the world's championship rodeo. I noticed an army cot in his dressing-room. "I lie down there every afternoon," Gene Autry said, "and get an hour's nap between performances. When I am making pictures in Hollywood," he continued, "I often relax in a big easy chair and get two or three ten-minute naps a day. They buck me up tremendously."

Edison attributed his enormous energy and endurance to his habit of sleeping whenever he wanted to.

I interviewed Henry Ford shortly before his eightieth birthday. I was surprised to see how fresh and fine he looked. I asked him the secret. He said: "I never stand up when I can sit down; and I never sit down when I can lie down."

Horace Mann, "the father of modern education", did the same thing as he grew older. When he was president of Antioch College, he used to stretch out on a couch while interviewing students.

I persuaded a motion-picture director in Hollywood to try a similar technique. He confessed that it worked miracles. I refer to Jack Chertock, who is now one of Metro-Goldwyn-Mayer's top directors. When he came to see me a few years ago, he was then head of the short-feature department of M-G-M. Worn out and exhausted, he had tried everything: tonics, vitamins, medicine. Nothing helped much. I suggested that he take a vacation every day. How? By stretching out in his office and relaxing while holding conferences with his staff writers.

When I saw him again, two years later, he said: "A miracle has happened. That is what my own physicians call it. I used to sit up in my chair, tense and taut, while discussing ideas for our short features. Now I stretch out on the office couch during these conferences. I feel better than I have felt in twenty years. Work two hours a day longer, yet I rarely get tired."

How does all this apply to you? If you are a stenographer, you can't take naps in the office as Edison did, and as Sam Goldwyn does; and if you are an accountant, you can't stretch out on the couch while discussing a financial statement with the boss. But if you live in a small city and go home for lunch, you may be able to take a ten-minute nap after lunch. That is what General George C. Marshall used to do. He felt he was so busy directing the U.S. Army in wartime that he had to rest at noon. If you are over fifty and feel you are too rushed to do it, then buy immediately all the life insurance you can get. Funerals come high-and suddenly-these days; and the little woman may want to take your insurance money and marry a younger man!

If you can't take a nap at noon, you can at least try to lie down for an hour before the evening meal. It is cheaper than a highball; and, over a long stretch, it is 5,467 times more effective. If you can sleep for an hour around five, six, or seven o'clock, you can add one hour a day to your waking life. Why? How? Because an hour's nap before the evening meal plus six hours' sleep at night-a total of seven hours-will do you more good than eight hours of unbroken sleep.

A physical worker can do more work if he takes more time out for rest. Frederick Taylor demonstrated that while working as a scientific management engineer with the Bethlehem Steel Company. He observed that labouring men were loading approximately 12 1/2 tons of pig-iron per man each day on freight cars and that they were exhausted at noon. He made a scientific study of all the fatigue factors involved, and declared that these men should be loading not 12 1/2 tons of pig-iron per day, but fortyseven tons per day! He figured that they ought to do almost four times as much as they were doing, and not be exhausted. But prove it!

Taylor selected a Mr. Schmidt who was required to work by the stop-watch. Schmidt was told by the man who stood over him with a watch: "Now pick up a 'pig' and walk. ... Now sit down and rest. ... Now walk. ... Now rest."

What happened? Schmidt carried forty-seven tons of pig-iron each day while the other men carried only 12 1/2 tons per man. And he practically never failed to work at this pace during the three years that Frederick Taylor was at Bethlehem. Schmidt was able to do this because he rested before he got tired. He worked approximately 26 minutes out of the hour and rested 34 minutes. He rested more than he worked-yet he did almost four times as much work as the others! Is this mere hearsay? No, you can read the record yourself in Principles of Scientific Management by Frederick Winslow Taylor.

Let me repeat: do what the Army does-take frequent rests. Do what your heart does-rest before you get tired, and you will add one hour a day to your waking life.

✦ ✦ ✦

Chapter 24

What Makes You Tired-and What You Can Do About It

Here is an astounding and significant fact: Mental work alone can't make you tired. Sounds absurd. But a few years ago, scientists tried to find out how long the human brain could labour without reaching "a diminished capacity for work", the scientific definition of fatigue. To the amazement of these scientists, they discovered that blood passing through the brain, when it is active, shows no fatigue at all! If you took blood from the veins of a day labourer while he was working, you would find it full of "fatigue toxins" and fatigue products. But if you took a drop of blood from the brain of an Albert Einstein, it would show no fatigue toxins whatever at the end of the day.

So far as the brain is concerned, it can work "as well and as swiftly at the end of eight or even twelve hours of effort as at the beginning". The brain is utterly tireless. ... So what makes you tired? Psychiatrists declare that most of our fatigue derives from our mental and emotional attitudes. One of England's most distinguished psychiatrists, J.A. Hadfield, says in his book *The Psychology of Power*: "the greater part of the fatigue from which we suffer is of mental origin; in fact exhaustion of purely physical origin is rare."

One of America's most distinguished psychiatrists, Dr. A.A. Brill, goes even further. He declares: "One hundred per cent of the fatigue of the sedentary worker in good health is due to psychological factors, by which we mean emotional factors."

What kinds of emotional factors tire the sedentary (or sitting) worker? Joy? Contentment? No! Never! Boredom, resentment, a feeling of not being appreciated, a feeling of futility, hurry, anxiety, worry—those are the emotional factors that exhaust the sitting worker, make him susceptible to colds, reduce his output, and send him home with a nervous headache. Yes, we get tired because our emotions produce nervous tensions in the body.

The Metropolitan Life Insurance Company pointed that out in a leaflet on fatigue: "Hard work by itself," says this great life-insurance company, "seldom causes fatigue which cannot be cured by a good sleep or rest. ... Worry, tenseness, and emotional upsets are three of the biggest causes of fatigue. Often they are to blame when physical or mental work seems to be the cause. ... Remember that a tense muscle is a working muscle. Ease up! Save energy for important duties."

Stop now, right where you are, and give yourself a check-up. As you read these lines, are you scowling at the book? Do you feel a strain between the eyes? Are you sitting relaxed in your chair? Or are you hunching up your shoulders? Are the muscles of your face tense? Unless your entire body is as limp and relaxed as an old rag doll, you are at this very moment producing nervous tensions and muscular tensions. You are producing nervous tensions and nervous fatigue!

Why do we produce these unnecessary tensions in doing mental work? Josselyn says: "I find that the chief obstacle ... is the almost universal belief that hard work requires a feeling of effort, else it is not well done." So we scowl when we concentrate. We hunch up our shoulders. We call on our muscles to make the motion of effort, which in no way assists our brain in its work.

Here is an astonishing and tragic truth: millions of people who wouldn't dream of wasting dollars go right on wasting and squandering their energy with the recklessness of seven drunken sailors in Singapore.

What is the answer to this nervous fatigue? Relax! Relax! Relax! Learn to relax while you are doing your work!

Easy? No. You will probably have to reverse the habits of a lifetime. But it is worth the effort, for it may revolutionise your life! William James said, in his essay "The Gospel of Relaxation": "The American over-tension and jerkiness and breathlessness and intensity and agony of expression ... are bad habits, nothing more or less." Tension is a habit. Relaxing is a habit. And bad habits can be broken, good habits formed.

How do you relax? Do you start with your mind, or do you start with your nerves? You don't start with either. You always begin to relax with your muscles!

Let's give it a try. To show how it is done, suppose we start with your eyes. Read this paragraph through, and when you've reached the end, lean back, close your eyes, and say to your eyes silently: "Let go. Let go. Stop straining, stop frowning. Let go. Let go." Repeat that over and over very slowly for a minute ....

Didn't you notice that after a few seconds the muscles of the eyes began to obey? Didn't you feel as though some hand had wiped away the tension? Well, incredible as it seems, you have sampled in that one minute the whole key and secret to the art of relaxing. You can do the same thing with the jaw, with the muscles of the face, with the neck, with the shoulders, the whole of the body. But the most important organ of all is the eye. Dr. Edmund Jacobson of the University of Chicago has gone so far as to say that if you can completely relax the muscles of the eyes, you can forget all your troubles! The reason the eyes are so important in relieving nervous tension is that they burn up one-fourth of all the nervous energies consumed by the body. That is also why so many people with perfectly sound vision suffer from "eyestrain". They are tensing the eyes.

Vicki Baum, the famous novelist, says that when she was a child, she met an old man who taught her one of the most important lessons she ever learned. She had fallen down and cut her knees and hurt her wrist. The old man picked her up; he had once been a circus clown; and, as he brushed her off, he said: "The reason you injured yourself was because you don't know how to relax. You have to pretend you are as limp as a sock, as an old crumpled sock. Come, I'll show you how to do it."

That old man taught Vicki Baum and the other children how to fall, how to do flip-flops, and how to turn somersaults. And always he insisted: "Think of yourself as an old crumpled sock. Then you've got to relax!"

You can relax in odd moments, almost anywhere you are. Only don't make an effort to relax. Relaxation is the absence of all tension and effort. Think ease and relaxation. Begin by thinking relaxation of the muscles of your eyes and your face, saying over and over: "Let go ... let go ... let go and relax." Feel the energy flowing out of your facial muscles to the centre of your body. Think of yourself as free from tension as a baby.

That is what Galli-Curci, the great soprano, used to do. Helen Jepson told me that she used to see Galli- Curci before a performance, sitting in a chair with all her muscles relaxed and her lower jaw so limp it actually sagged. An excellent practice-it kept her from becoming too nervous before her stage entrance; it prevented fatigue.

Here are five suggestions that will help you learn to relax:

1 Read one of the best books ever written on this subject: *Release from Nervous Tension*, by Dr. David Harold Fink.

2 Relax in odd moments. Let your body go limp like an old sock. I keep an old, maroon-coloured sock on my desk as I work-keep it there as a reminder of how limp I ought to be. If you haven't got a sock, a cat will do. Did you ever pick up a kitten sleeping in the sunshine? If so, both ends sagged like a wet newspaper. Even the yogis in India say that if you want to master the art of relaxation, study the cat. I never saw a tired cat, a cat with a nervous breakdown, or a cat suffering from insomnia, worry, or stomach ulcers. You will probably avoid these disasters if you learn to relax as the cat does.

3 Work, as much as possible, in a comfortable position. Remember that tensions in the body produce aching shoulders and nervous fatigue.

4 Check yourself four or five times a day, and say to yourself: "Am I making my work harder than it actually is? Am I using muscles that have nothing to do with the work I am doing?" This will help you form the habit of relaxing, and as Dr. David Harold Fink says: "Among those who know psychology best, it is habits two to one."

5 Test yourself again at the end of the day, by asking yourself: "Just how tired am I? If I am tired, it is not because of the mental work I have done but because of the way I have done it." "I measure my accomplishments," says Daniel W. Josselyn, "not by how tired I am at the end of the day, but how tired I am not." He says: "When I feel particularly tired at the end of the day, or when irritability proves that my nerves are tired, I know beyond question that it has been an inefficient day both as to quantity and quality." If every businessman would learn that same lesson, the death rate from "hypertension" diseases would drop overnight. And we would stop filling up our sanatoriums and asylums with men who have been broken by fatigue and worry.

⁂

Chapter 25

How The Housewife Can Avoid Fatigue and Keep Looking Young

One day last autumn, my associate flew up to Boston to attend a session of one of the most unusual medical classes in the world. Medical? Well, yes, it meets once a week at the Boston Dispensary, and the patients who attend it get regular and thorough medical examinations before they are admitted. But actually this class is a psychological clinic. Although it is officially called the Class in Applied Psychology (formerly the Thought Control Class-a name suggested by the first member), its real purpose is to deal with people who are ill from worry. And many of these patients are emotionally disturbed housewives.

How did such a class for worriers get started? Well, in 1930, Dr. Joseph H. Pratt-who, by the way, had been a pupil of Sir William Osier-observed that many of the outpatients who came to the Boston Dispensary apparently had nothing wrong with them at all physically; yet they had practically all the symptoms that flesh is heir to. One woman's hands were so crippled with "arthritis" that she had lost all use of them. Another was in agony with all the excruciating symptoms of "cancer of the stomach". Others had backaches, headaches, were chronically tired, or had vague aches and pains. They actually felt these pains. But the most exhaustive medical examinations showed that nothing whatever was wrong with these women-in the physical sense. Many old-fashioned doctors would have said it was all imagination- "all in the mind".

But Dr. Pratt realised that it was no use to tell these patients to "go home and forget it". He knew that most of these women didn't want to be sick; if it was so easy to forget their ailments, they would do so themselves. So what could be done?

He opened his class-to a chorus of doubts from the medical doubters on the sidelines. And the class worked wonders! In the eighteen years that have passed since it started, thousands of patients have been "cured" by attending it. Some of the patients have been coming for years-as religious in their attendance as though going to church. My assistant talked to a woman who had hardly missed a session in more than nine years. She said that when she first went to the clinic, she was thoroughly convinced she had a floating kidney and some kind of heart ailment. She was so worried and tense that she occasionally lost her eyesight and had spells of blindness. Yet today she is confident and cheerful and in excellent health. She looked only about forty, yet she held one of her grandchildren asleep in her lap. "I used to worry so much about my family troubles," she said, "that I wished I could die. But I learned at this clinic the futility of worrying. I learned to stop it. And I can honestly say now that my life is serene."

Dr. Rose Hilferding, the medical adviser of the class, said that she thought one of the best remedies for lightening worry is "talking your troubles over with someone you trust. We call it catharsis," she said. "When patients come here, they can talk their troubles over at length, until they get them off their minds. Brooding over worries alone, and keeping them to oneself, causes great nervous tension. We all have to share our troubles. We have to share worry. We have to feel there is someone in the world who is willing to listen and able to understand."

My assistant witnessed the great relief that came to one woman from talking out her worries. She had domestic worries, and when she first began to talk, she was like a wound-up spring. Then gradually, as she kept on talking, she began to calm down. At the end of the interview, she was actually smiling. Had the problem been solved? No, it wasn't that easy. What caused the change was talking to someone, getting a little advice and a little human sympathy. What had really worked the change was the tremendous healing value that lies in-words!

Psycho-analysis is based, to some extent, on this healing power of words. Ever since the days of Freud, analysts have known that a patient could find relief from his inner anxieties if he could talk, just talk. Why is this so? Maybe because by talking, we gain a little better insight into our troubles, get a better perspective. No one knows the whole answer. But all of us know that "spitting it out" or "getting it off our chests" bring almost instant relief.

So the next time we have an emotional problem, why don't we look around for someone to talk to? I don't mean, of course, to go around making pests of ourselves by whining and complaining to everyone in sight. Let's decide on someone we can trust, and make an appointment. Maybe a relative, a doctor, a lawyer, a minister, or priest. Then say to that person: "I want your advice. I have a problem, and I wish you would listen while I put it in words. You may be able to advise me. You may see angles to this thing that I can't see myself. But even if you can't, you will help me tremendously if you will just sit and listen while I talk it out."

However, if you honestly feel that there is no one you can talk to, then let me tell you about the Save-a-Life League- it has no connection with the Boston Dispensary. The Save-a-Life League is one of the most unusual leagues in the world. It was originally formed to save possible suicides. But as the years went on, it expanded its scope to give spiritual counsel to those who are unhappy and in emotional need. I talked for some time to Miss Lona B. Bonnell, who interviews people who come for advice to the Save-a-Life League. She told me that she would be glad to answer letters from readers of this book. If you write to the Save-a-Life League, 505 Fifth Avenue, New York City, your letter and your troubles will be held in strictest confidence. Frankly, I would advise you to go to someone you can talk to in person if you can, for that will give you greater relief. But if that is out of the question, then why not write to this league?

Talking things out, then, is one of the principle therapies used at the Boston Dispensary Class. But here are some other ideas we picked up at class—things you, as a housewife, can do in your home.

1. Keep a notebook or scrapbook for "inspirational" reading. Into this book you can paste all the poems, or short prayers, or quotations, which appeal to you personally and give you a lift. Then, when a rainy afternoon sends your spirits plunging down, perhaps you can find a recipe in this book for dispelling the gloom. Many patients at the Dispensary have kept such notebooks for years. They say it is a spiritual "shot in the arm".

2. Don't dwell too long on the shortcomings of others! Sure, your husband has faults! If he had been a saint, he never would have married you. Right? One woman at the class who found herself developing into a scolding, nagging, and haggard-faced wife, was brought up short with the question: "What would you do if your husband died?" She was so shocked by the idea that she immediately sat down and drew up a list of all her husband's good points. She made quite a list. Why don't you try the same thing the next time you feel you married a tight-fisted tyrant? Maybe you'll find, after reading his virtues, that he's a man you'd like to meet!

3. Get interested in your neighbours! Develop a friendly, healthy interest in the people who share the life on your street. One ailing woman who felt herself so "exclusive" that she hadn't any friends, was told to try to make up a story about the next person she met. She began, in the street-car, to weave backgrounds and settings for the people she saw. She tried to imagine what their lives had been like. First thing you know, she was talking to people everywhere—and today she is happy, alert, and a charming human being cured of her "pains".

4. Make up a schedule for tomorrow's work before you go to bed tonight. The class found that many wives feel driven and harassed by the unending round of housework and things they must do. They never got their work finished. They were chased by the clock. To cure this sense of hurry, and worry, the suggestion was made that they draw up a schedule each night for the following day. What happened? More work accomplished; much less fatigue; a feeling of pride and achievement; and time left over to rest and enjoyment.

5. Finally, avoid tension and fatigue. Relax! Relax! Nothing will make you look old sooner than tension and fatigue. Nothing will work such havoc with your freshness and looks! My assistant sat for an hour in the Boston Thought Control Class, while Professor Paul E. Johnson, the director, went over many of the principles we have already discussed in the previous chapter—the rules for relaxing. At the end of ten minutes of these relaxing exercises, which my assistant did with the others, she was almost asleep sitting upright in her chair! Why is such stress laid on this physical relaxing? Because the clinic knows-as other doctors know-that if you're going to get the worry-kinks out of people, they've got to relax!

Yes, you, as a housewife, have got to relax! You have one great advantage—you can lie down whenever you want to, and you can lie on the floor! Strangely enough, a good hard floor is better to relax on than an inner-spring bed. It gives more resistance. It is good for the spine.

All right, then, here are some exercises you can do in your home. Try them for a week—and see what you do for your looks and disposition!

a. Lie flat on the floor whenever you feel tired. Stretch as tall as you can. Roll around if you want to. Do it twice a day.

b. Close your eyes. You might try saying, as Professor Johnson recommended, something like this: ' 'The sun is shining overhead. The sky is blue and sparkling. Nature is calm and in control of the world and I, as nature's child, am in tune with the Universe." Or—better still—pray!

c. If you cannot lie down, because the roast is in the oven and you can't spare the time, then you can achieve almost the same effect sitting down in a chair. A hard, upright chair is the best for relaxing. Sit upright in the chair like a seated Egyptian statue, and let your hands rest, palms down, on the tops of your thighs.

d. Now, slowly tense the toes-then let them relax. Tense the muscles in your legs—and let them relax. Do this slowly upward, with all the muscles of your body, until you get to the neck. Then let your head roll around heavily, as though it were a football. Keep saying to your muscles (as in the previous chapter): "Let go ... let go ..."

e. Quiet your nerves with slow, steady breathing. Breathe from deep down. The yogis of India were right: rhythmical breathing is one of the best methods ever discovered for soothing the nerves.

f. Think of the wrinkles and frowns in your face, and smooth them all out. Loosen up the worry-creases you feel between your brows, and at the sides of your mouth. Do this twice a day, and maybe you won't have to go to a beauty parlour to get a massage. Maybe the lines will disappear from the inside out!

+ + +

Chapter 26

Four Good Working Habits That Will Help Prevent Fatigue And Worry

Good Working Habit No. 1: Clear Your Desk of All Papers Except Those Relating to the Immediate Problem at Hand. Roland L. Williams, President of Chicago and North-western Railway, says: "A person with his desk piled high with papers on various matters will find his work much easier and more accurate if he clears that desk of all but the immediate problem on hand. I call this good housekeeping, and it is the number one step towards efficiency."

If you visit the Library of Congress in Washington, D.C., you will find five words painted on the ceiling-five words written by the poet Pope:

"Order is Heaven's first law."

Order ought to be the first law of business, too. But is it? No, the average businessman's desk is cluttered up with papers that he hasn't looked at for weeks. In fact, the publisher of a New Orleans newspaper once told me that his secretary cleared up one of his desks and found a typewriter that had been missing for two years!

The mere sight of a desk littered with unanswered mail and reports and memos is enough to breed confusion, tension, and worries. It is much worse than that. The constant reminder of "a million things to do and no time to do them" can worry you not only into tension and fatigue, but it can also worry you into high blood pressure, heart trouble, and stomach ulcers.

Dr. John H. Stokes, professor, Graduate School of Medicine, University of Pennsylvania, read a paper before the National Convention of the American Medical Association—a paper entitled "Functional Neuroses as Complications of Organic Disease". In that paper, Dr. Stokes listed eleven conditions under the title: "What to Look for in the Patient's State of Mind". Here is the first item on that list:

"The sense of must or obligation; the unending stretch of things ahead that simply have to be done."

But how can such an elementary procedure as clearing your desk and making decisions help you avoid this high pressure, this sense of must, this sense of an "unending stretch of things ahead that simply have to be done"? Dr. William L. Sadler, the famous psychiatrist, tells of a patient who, by using this simple device, avoided a nervous breakdown. The man was an executive in a big Chicago firm. When he came to Dr. Sadler's office, he was tense, nervous, worried. He knew he was heading for a tailspin, but he couldn't quit work. He had to have help.

"While this man was telling me his story," Dr. Sadler says, "my telephone rang. It was the hospital calling; and, instead of deferring the matter, I took time right then to come to a decision. I always settle questions, if possible, right on the spot. I had no sooner hung up than the phone rang again. Again an urgent matter, which I took time to discuss. The third interruption came when a colleague of mine came to my office for advice on a patient who was critically ill. When I had finished with him, I turned to my caller and began to apologise for keeping him waiting. But he had brightened up. He had a completely different look on his face."

"Don't apologise, doctor!" this man said to Sadler. "In the last ten minutes, I think I've got a hunch as to what is wrong with me. I'm going back to my offices and revise my working habits .... But before I go, do you mind if I take a look in your desk?"

Dr. Sadler opened up the drawers of his desk. All empty-except for supplies. "Tell me," said the patient, "where do you keep your unfinished business?"

"Finished!" said Sadler.

"And where do you keep your unanswered mail?"

"Answered!" Sadler told him. "My rule is never to lay down a letter until I have answered it. I dictate the reply to my secretary at once."

Six weeks later, this same executive invited Dr. Sadler to come to his office. He was changed—and so was his desk. He opened the desk drawers to show there was no unfinished business inside of the desk. "Six weeks ago," this executive said, "I had three different desks in two different offices-and was snowed under by my work. I was never finished. After talking to you, I came back here and cleared out a wagon-load of reports and old papers. Now I work at one desk, settle things as they come up, and don't have a mountain of unfinished business nagging at me and making me tense and worried. But the most astonishing thing is I've recovered completely. There is nothing wrong any more with my health!"

Charles Evans Hughes, former Chief Justice of the United States Supreme Court, said: "Men do not die from overwork. They die from dissipation and worry." Yes, from dissipation of their energies—and worry because they never seem to get their work done.

Good Working Habit No. 2: Do Things in the Order of Their Importance.

Henry L. Dougherty, founder of the nation-wide Cities Service Company, said that regardless of how much salary he paid, there were two abilities he found it almost impossible to find.

Those two priceless abilities are: first, the ability to think. Second, the ability to do things in the order of their importance.

Charles Luckman, the lad who started from scratch and climbed in twelve years to president of the Pepsodent Company, got a salary of a hundred thousand dollars a year, and made a million dollars besides—that lad declares that he owes much of his success to developing the two abilities that Henry L. Dougherty said he found almost impossible to find. Charles Luckman said: "As far back as I can remember, I have got up at five o'clock in the morning because I can think better then than any other time—I can think better then and plan my day, plan to do things in the order of their importance." Franklin Bettger, one of America's most successful insurance salesmen, doesn't wait until five o'clock in the morning to plan his day. He plans it the night before-sets a goal for himself- a goal to sell a certain amount of insurance that day. If he fails, that amount is added to the next day-and so on.

I know from long experience that one is not always able to do things in the order of their importance, but I also know that some kind of plan to do first things first is infinitely better than extemporising as you go along.

If George Bernard Shaw had not made it a rigid rule to do first things first, he would probably have failed as a writer and might have remained a bank cashier all his life. His plan called for writing five pages each day. That plan and his dogged determination to carry it through saved him. That plan inspired him to go right on writing five pages a day for nine heartbreaking years, even though he made a total of only thirty dollars in those nine years-about a penny a day.

Good Working Habit No. 3. When You Face a Problem, Solve It Then and There if You Have the Facts Necessary to Make a Decision. Don't Keep Putting off Decisions.

One of my former students, the late H.P. Howell, told me that when he was a member of the board of directors of U.S. Steel, the meetings of the board were often long-drawn-out affairs-many problems were discussed, few decisions were made. The result: each member of the board had to carry home bundles of reports to study.

Finally, Mr. Howell persuaded the board of directors to take up one problem at a time and come to a decision. No procrastination-no putting off. The decision might be to ask for additional facts; it might be to do something or do nothing. But a decision was reached on each problem before passing on to the next. Mr. Howell told me that the results were striking and salutary: the docket was cleared. The calendar was clean. No longer was it necessary for each member to carry home a bundle of reports. No longer was there a worried sense of unresolved problems.

A good rule, not only for the board of directors of U.S. Steel, but for you and me.

Good Working Habit No. 4: Learn to Organise, Deputise, and Supervise.

Many a businessman is driving himself to a premature grave because he has never learned to delegate responsibility to others, insists on doing everything himself. Result: details and confusion overwhelm him. He is driven by a sense of hurry, worry, anxiety, and tension. It is hard to learn to delegate responsibilities. I know. It was hard for me, awfully hard. I also know from experience the disasters that can be caused by delegating authority to the wrong people. But difficult as it is to delegate authority, the executive must do it if he is to avoid worry, tension, and fatigue.

The man who builds up a big business, and doesn't learn to organise, deputise, and supervise, usually pops off with heart trouble in his fifties or early sixties—heart trouble caused by tension and worries. Want a specific instance? Look at the death notices in your local paper.

✦ ✦ ✦

Chapter 27

How To Banish The Boredom That Produces Fatigue, Worry, And Resentment

One of the chief causes of fatigue is boredom. To illustrate, let's take the case of Alice, a stenographer who lives on your street. Alice came home one night utterly exhausted. She acted fatigued. She was fatigued. She had a headache. She had a backache. She was so exhausted she wanted to go to bed without waiting for dinner. Her mother pleaded … . She sat down at the table. The telephone rang. The boy friend! An invitation to a dance! Her eyes sparkled. Her spirits soared. She rushed upstairs, put on her Alice-blue gown, and danced until three o'clock in the morning; and when she finally did get home, she was not the slightest bit exhausted. She was, in fact, so exhilarated she couldn't fall asleep.

Was Alice really and honestly tired eight hours earlier, when she looked and acted exhausted? Sure she was. She was exhausted because she was bored with her work, perhaps bored with life. There are millions of Alices. You may be one of them.

It is a well-known fact that your emotional attitude usually has far more to do with producing fatigue than has physical exertion. A few years ago, Joseph E. Barmack, Ph.D., published in the Archives of Psychology a report of some of his experiments showing how boredom produces fatigue. Dr. Barmack put a group of students through a series of tests in which, he knew, they could have little interest. The result? The students felt tired and sleepy, complained of headaches and eyestrain, felt irritable. In some cases, even their stomachs were upset. Was it all "imagination"? No. Metabolism tests were taken of these students.

These tests showed that the blood pressure of the body and the consumption of oxygen actually decrease when a person is bored, and that the whole metabolism picks up immediately as soon as he begins to feel interest and pleasure in his work!

We rarely get tired when we are doing something interesting and exciting. For example, I recently took a vacation in the Canadian Rockies up around Lake Louise. I spent several days trout fishing along Corral Creek, fighting my way through brush higher than my head, stumbling over logs, struggling through fallen timber—yet after eight hours of this, I was not exhausted. Why? Because I was excited, exhilarated. I had a sense of high achievement: six cut-throat trout. But suppose I had been bored by fishing, then how do you think I would have felt? I would have been worn out by such strenuous work at an altitude of seven thousand feet.

Even in such exhausting activities as mountain climbing, boredom may tire you far more than the strenuous work involved. For example, Mr. S. H. Kingman, president of the Farmers and Mechanics Savings Bank of Minneapolis, told me of an incident that is a perfect illustration of that statement. In July, 1943, the Canadian government asked the Canadian Alpine Club to furnish guides to train the members of the Prince of Wales Rangers in mountain climbing. Mr. Kingman was one of the guides chosen to train these soldiers. He told me how he and the other guides—men ranging from forty-two to fifty-nine years of age—took these young army men on long hikes across glaciers and snow fields and up a sheer cliff of forty feet, where they had to climb with ropes and tiny foot-holds and precarious hand-holds. They climbed Michael's Peak, the Vice-President Peak, and other unnamed peaks in the Little Yoho Valley in the Canadian Rockies. After fifteen hours of mountain climbing, these young men, who were in the pink of condition (they had just finished a six-week course in tough Commando training), were utterly exhausted.

Was their fatigue caused by using muscles that had not been hardened by Commando training? Any man who had ever been through Commando training would hoot at such a ridiculous question! No, they were utterly exhausted because they were bored by mountain climbing. They were so tarred, that many of them fell asleep without waiting to eat. But the guides—men who were two and three times as old as the soldiers—were they tired? Yes, but not exhausted. The guides ate dinner and stayed up for hours, talking about the day's experiences. They were not exhausted because they were interested. When Dr. Edward Thorndike of Columbia was conducting experiments in fatigue, he kept young men awake for almost a week by keeping them constantly interested. After much investigation, Dr. Thorndike is reported to have said: "Boredom is the only real cause of diminution of work."

If you are a mental worker, it is seldom the amount of work you do that makes you tired. You may be tired by the amount of work you do not do. For example, remember the day last week when you were constantly interrupted. No letters answered. Appointments broken. Trouble here and there. Everything went wrong that day. You accomplished nothing whatever, yet you went home exhausted-and with a splitting head.

The next day everything clicked at the office. You accomplished forty times more than you did the previous day. Yet you went home fresh as a snowy-white gardenia. You have had that experience. So have I.

The lesson to be learned? Just this: our fatigue is often caused not by work, but by worry, frustration, and resentment.

While writing this chapter, I went to see a revival of Jerome Kern's delightful musical comedy, Show Boat. Captain Andy, captain of the Cotton Blossom, says, in one of his philosophical interludes: "The lucky folks are the ones that get to do the things they enjoy doing." Such folks are lucky because they have more energy, more happiness, less worry, and less fatigue. Where your interests are, there is your energy also. Walking ten blocks with a nagging wife can be more fatiguing than walking ten miles with an adoring sweetheart.

And so what? What can you do about it? Well, here is what one stenographer did about it—a stenographer working for an oil company in Tulsa, Oklahoma. For several days each month, she had one of the dullest jobs imaginable: filling out printed forms for oil leases, inserting figures and statistics. This task was so boring that she resolved, in self-defence, to make it interesting. How? She had a daily contest with herself. She counted the number of forms she filled out each morning, and then tried to excel that record in the afternoon. She counted each day's total and tried to better it the next day. Result? She was soon able to fill out more of these dull printed forms than any other stenographer in her division. And what did all this get her? Praise? No. ... Thanks? No. ... Promotion? No. ... Increased pay? No. ... But it did help to prevent the fatigue that is spawned by boredom. It did give her a mental stimulant. Because she had done her best to make a dull job interesting, she had more energy, more zest, and got far more happiness out of her leisure hours. I happen to know this story is true, because I married that girl.

Here is the story of another stenographer who found it paid to act as if her work were interesting. She used to fight her work. But no more. Her name is Miss Vallie G. Golden, and she lives at 473 South Kenilworth Avenue, Elmhurst, Illinois. Here is her story, as she wrote it to me:

"There are four stenographers in my office and each of us is assigned to take letters from several men. Once in a while we get jammed up in these assignments; and one day, when an assistant department head insisted that I do a long letter over, I started to rebel. I tried to point out to him that the letter could be corrected without being retyped—and he retorted that if I didn't do it over, he would find someone else who would! I was absolutely fuming! But as I started to retype this letter, it suddenly occurred to me that there were a lot of other people who would jump at the chance to do the work I was doing. Also, that I was being paid a salary to do just that work. I began to feel better. I suddenly made up my mind to do my work as if I actually enjoyed it—even though I despised it. Then I made this important discovery: if I do my work as if I really enjoy it, then I do enjoy it to some extent I also found I can work faster when I enjoy my work. So there is seldom any need now for me to work overtime. This new attitude of mine gained me the reputation of being a good worker. And when one of the department superintendents needed a private secretary, he asked for me for the job- because, he said, I was willing to do extra work without being sulky! This matter of the power of a changed mental attitude," wrote Miss Golden, "has been a tremendously important discovery to me. It has worked wonders!"

Without perhaps being conscious of it. Miss Vallie Golden was using the famous "as if" philosophy. William James counseled us to act "as if" we were brave, and we would be brave; and to act "as if" we were happy, and we would be happy, and so on.

Act "as if" you were interested in your job, and that bit of acting will tend to make your interest real. It will also tend to decrease your fatigue, your tensions, and your worries.

A few years ago, Harlan A. Howard made a decision that completely altered his life. He resolved to make a dull job interesting—and he certainly had a dull one: washing plates, scrubbing counters, and dishing out ice-cream in the high-school lunchroom while the other boys were playing ball or kidding the girls. Harlan Howard despised his job-but since he had to stick to it, he resolved to study ice-cream, how it was made, what ingredients were used, why some ice-creams were better than others. He studied the chemistry of ice-cream, and became a whiz in the high-school chemistry course. He was so interested now in food chemistry that he entered the Massachusetts State College and majored in the field of "food technology". When the New York Cocoa Exchange offered a hundred-dollar prize for the best paper on uses of cocoa and chocolate-a prize open to all college students-who do you suppose won it? ... That's right. Harlan Howard.

When he found it difficult to get a job, he opened a private laboratory in the basement of his home at 750 North Pleasant Street, Amherst, Massachusetts. Shortly after that, a new law was passed. The bacteria in milk had to be counted. Harlan A. Howard was soon counting bacteria for the fourteen milk companies in Amherst-and he had to hire two assistants.

Where will he be twenty-five years from now? Well, the men who are now running the business of food chemistry will be retired then, or dead; and their places will be taken by young lads who are now radiating initiative and enthusiasm. Twenty-five years from now, Harlan A. Howard will probably be one of the leaders in his profession, while some of his class-mates to whom he used to sell ice-cream over the counter will be sour, unemployed, cursing the government, and complaining that they never had a chance. Harlan A. Howard might never have had a chance, either, if he hadn't resolved to make a dull job interesting.

Years ago, there was another young man who was bored with his dull job of standing at a lathe, turning out bolts in a factory. His first name was Sam. Sam wanted to quit, but he was afraid he couldn't find another job. Since he had to do this dull work, Sam decided he would make it interesting. So he ran a race with the mechanic operating a machine beside him. One of them was to trim off the rough surfaces on his machine, and the other was to trim the bolts down to the proper diameter. They would switch machines occasionally and see who could turn out the most bolts. The foreman, impressed with Sam's speed and accuracy, soon gave him a better job. That was the start of a whole series of promotions. Thirty years later, Sam—Samuel Vauclain—was president of the Baldwin Locomotive Works. But he might have remained a mechanic all his life if he had not resolved to make a dull job interesting.

H. V. Kaltenborn-the famous radio news analyst, once told me how he made a dull job interesting. When he was twenty-two years old, he worked his way across the Atlantic on a cattle boat, feeding and watering the steers. After making a bicycle tour of England, he arrived in Paris, hungry and broke. Pawning his camera for five dollars, he put an ad. in the Paris edition of The New York Herald and got a job selling steropticon machines. If you are forty years old, you may remember those old-fashioned stereoscopes that we used to hold up before our eyes to look at two pictures exactly alike. As we looked, a miracle happened. The two lenses in the stereoscope transformed the two pictures into a single scene with the effect of a third dimension. We saw distance. We got an astounding sense of perspective.

Well, as I was saying, Kaltenborn started out selling these machines from door to door in Paris—and he couldn't speak French. But he earned five thousand dollars in commissions the first year, and made himself one of the highest-paid salesmen in France that year. H.V. Kaltenborn told me that this experience did as much to develop within him the qualities that make for success as did any single year of study at Harvard. Confidence? He told me himself that after that experience, he felt he could have sold The Congressional Record to French housewives.

That experience gave him an intimate understanding of French life that later proved invaluable in interpreting, on the radio, European events.

How did he manage to become an expert salesman when he couldn't speak French? Well, he had his employer write out his sales talk in perfect French, and he memorised it. He would ring a door-bell, a housewife would answer, and Kaltenborn would begin repeating his memorised sales talk with an accent so terrible it was funny. He would show the housewife his pictures, and when she asked a question, he would shrug his shoulders and say: "An American ... an American." He would then take off his hat and point to a copy of the sales talk in perfect French that he had pasted in the top of his hat. The housewife would laugh, he would laugh-and show her more pictures. When H. V. Kaltenborn told me about this, he confessed that the job had been far from easy. He told me that there was only one quality that pulled him through: his determination to make the job interesting. Every morning before he started out, he looked into the mirror and gave himself a pep talk: "Kaltenborn, you have to do this if you want to eat. Since you have to do it—why not have a good time doing it? Why not imagine every time you ring a door-bell that you are an actor before the footlights and that there's an audience out there looking at you. After all, what you are doing is just as funny as something on the stage. So why not put a lot of zest and enthusiasm into it?"

Mr. Kaltenborn told me that these daily pep talks helped him transform a task that he had once hated and dreaded into an adventure that he liked and made highly profitable.

When I asked Mr. Kaltenborn if he had any advice to give to the young men of America who are eager to succeed, he said: "Yes, go to bat with yourself every morning. We talk a lot about the importance of physical exercise to wake us up out of the half-sleep in which so many of us walk around. But we need, even more, some spiritual and mental exercises every morning to stir us into action. Give yourself a pep talk every day."

Is giving yourself a pep talk every day silly, superficial, childish? No, on the contrary, it is the very essence of sound psychology. "Our life is what our thoughts make it." Those words are just as true today as they were eighteen centuries ago when Marcus Aurelius first wrote them in his book of Meditations: "Our life is what our thoughts make it."

By talking to yourself every hour of the day, you can direct yourself to think thoughts of courage and happiness, thoughts of power and peace. By talking to yourself about the things you have to be grateful for, you can fill your mind with thoughts that soar and sing.

By thinking the right thoughts, you can make any job less distasteful. Your boss wants you to be interested in your job so that he will make more money. But let's forget about what the boss wants. Think only of what getting interested in your job will do for you. Remind yourself that it may double the amount of happiness you get out of life, for you spend about one half of your waking hours at your work, and if you don't find happiness in your work, you may never find it anywhere. Keep reminding yourself that getting interested in your job will take your mind off your worries, and, in the long run, will probably bring promotion and increased pay. Even if it doesn't do that, it will reduce fatigue to a minimum and help you enjoy your hours of leisure.

✦ ✦ ✦

Chapter 28

How To Keep From Worrying About Insomnia

Do you worry when you can't sleep well? Then it may interest you to know that Samuel Untermyer-the famous international lawyer-never got a decent night's sleep in his life.

When Sam Untermyer went to college, he worried about two afflictions-asthma and insomnia. He couldn't seem to cure either, so he decided to do the next best thing—take advantage of his wakefulness. Instead of tossing and turning and worrying himself into a breakdown, he would get up and study. The result? He began ticking off honours in all of his classes, and became one of the prodigies of the College of the City of New York.

Even after he started to practice law, his insomnia continued. But Untermyer didn't worry. "Nature," he said, "will take care of me." Nature did. In spite of the small amount of sleep he was getting, his health kept up and he was able to work as hard as any of the young lawyers of the New York Bar. He even worked harder, for he worked while they slept!

At the age of twenty-one, Sam Untermyer was earning seventy-five thousand dollars a year; and other young attorneys rushed to courtrooms to study his methods. In 1931, he was paid—for handling one case—what was probably the highest lawyer's fee in all history: a cool million dollars—cash on the barrelhead.

Still he had insomnia-read half the night-and then got up at five A.M. and started dictating letters. By the time most people were just starting work, his day's work would be almost half done. He lived to the age of eighty-one, this man who had rarely had a sound night's sleep; but if he had fretted and worried about his insomnia, he would probably have wrecked his life.

We spend a third of our lives sleeping, yet nobody knows what sleep really is. We know it is a habit and a state of rest in which nature knits up the ravelled sleeve of care, but we don't know how many hours of sleep each individual requires. We don't even know if we have to sleep at all!

Fantastic? Well, during the First World War, Paul Kern, a Hungarian soldier, was shot through the frontal lobe of his brain. He recovered from the wound, but curiously enough, couldn't fall asleep. No matter what the doctors did-and they tried all kinds of sedatives and narcotics, even hypnotism- Paul Kern couldn't be put to sleep or even made to feel drowsy.

The doctors said he wouldn't live long. But he fooled them. He got a job, and went on living in the best of health for years. He would lie down and close his eyes and rest, but he got no sleep whatever. His case was a medical mystery that upset many of our beliefs about sleep.

Some people require far more sleep than others. Toscanini needs only five hours a night, but Calvin Coolidge needed more than twice that much. Coolidge slept eleven hours out of every twenty-four. In other words, Toscanini has been sleeping away approximately one-fifth of his life, while Coolidge slept away almost half of his life.

Worrying about insomnia will hurt you far more than insomnia. For example, one of my students-Ira Sandner, of 173 Overpeck Avenue, Ridgefield Park, New Jersey-was driven nearly to suicide by chronic insomnia.

"I actually thought I was going insane," Ira Sandner told me. "The trouble was, in the beginning, that I was too sound a sleeper. I wouldn't wake up when the alarm clock went off, and the result was that I was getting to work late in the morning. I worried about it and, in fact, my boss warned me that I would have to get to work on time. I knew that if I kept on oversleeping, I would lose my job.

"I told my friends about it, and one of them suggested I concentrate hard on the alarm clock before I went to sleep. That started the insomnia! The tick-tick-tick of that blasted alarm clock became an obsession. It kept me awake, tossing, all night long! When morning came, I was almost ill. I was ill from fatigue and worry. This kept on for eight weeks. I can't put into words the tortures I suffered. I was convinced I was going insane. Sometimes I paced the floor for hours at a time, and I honestly considered jumping out of the window and ending the whole thing!

"At last I went to a doctor I had known all my life. He said: 'Ira, I can't help you. No one can help you, because you have brought this thing on yourself. Go to bed at night, and if you can't fall asleep, forget all about it. Just say to yourself: "I don't care a hang if I don't go to sleep. It's all right with me if I lie awake till morning." Keep your eyes closed and say: "As long as I just lie still and don't worry about it, I'll be getting rest, anyway."'

"I did that," says Sandner, "and in two weeks' time I was dropping off to sleep. In less than one month, I was sleeping eight hours, and my nerves were back to normal."

It wasn't insomnia that was killing Ira Sandner; it was his worry about it.

Dr. Nathaniel Kleitman, professor at the University of Chicago, has done more research work on sleep than has any other living man. He is the world's expert on sleep. He declares that he has never known anyone to die from insomnia. To be sure, a man might worry about insomnia until he lowered his vitality and was swept away by germs. But it was the worry that did the damage, not the insomnia itself.

Dr. Kleitman also says that the people who worry about insomnia usually sleep far more than they realise. The man who swears "I never slept a wink last night" may have slept for hours without knowing it. For example, one of the most profound thinkers of the nineteenth century, Herbert Spencer, was an old bachelor, lived in a boarding house, and bored everyone with his talk about his insomnia. He even put "stoppings" in his ears to keep out the noise and quiet his nerves. Sometimes he took opium to induce sleep. One night he and Professor Sayce of

Oxford shared the same room at a hotel. The next morning Spencer declared he hadn't slept a wink all night. In reality, it was Professor Sayce who hadn't slept a wink. He had been kept awake all night by Spencer's snoring.

The first requisite for a good night's sleep is a feeling of security. We need to feel that some power greater than ourselves will take care of us until morning. Dr. Thomas Hyslop, of the Great West Riding Asylum, stressed that point in an address before the British Medical Association. He said: "One of the best sleep-producing agents which my years of practice have revealed to me is prayer. I say this purely as a medical man. The exercise of prayer, in those who habitually exert it, must be regarded as the most adequate and normal of all the pacifiers of the mind and calmers of the nerves."

"Let God-and let go."

Jeanette MacDonald told me that when she was depressed and worried and had difficulty in going to sleep, she could always get "a feeling of security" by repeating Psalm XXII: "The Lord is my Shepherd; I shall not want. He maketh me to lie down in green pastures. He leadeth me beside the still waters. ..."

But if you are not religious, and have to do things the hard way, then learn to relax by physical measures. Dr. David Harold Fink, who wrote Release from Nervous Tension, says that the best way to do this is to talk to your body. According to Dr. Fink, words are the key to all kinds of hypnosis; and when you consistently can't sleep, it is because you have talked yourself into a case of insomnia. The way to undo this is to dehypnotise yourself—and you can do it by saying to the muscles of your body: "Let go, let go—loosen up and relax." We already know that the mind and nerves can't relax while the muscles are tense—so if we want to go to sleep, we start with the muscles. Dr. Fink recommends—and it works out in practice—that we put a pillow under the knees to ease the tension on the legs, and that we tuck small pillows under the arms for the very same reason. Then, by telling the jaw to relax, the eyes, the arms, and the legs, we finally drop off to sleep before we know what has hit us. I've tried it—I know. If you have trouble sleeping, get hold of Dr. Fink's book, *Release from Nervous Tension*, which I have mentioned earlier. It is the only book I know of that is both lively reading and a cure for insomnia.

One of the best cures for insomnia is making yourself physically tired by gardening, swimming, tennis, golf, skiing, or by just plain physically exhausting work. That is what Theodore Dreiser did. When he was a struggling young author, he was worried about insomnia, so he got a job working as a section hand on the New York Central Railway; and after a day of driving spikes and shoveling gravel, he was so exhausted that he could hardly stay awake long enough to eat.

If we get tired enough, nature will force us to sleep even while we are walking. To illustrate, when I was thirteen years old, my father shipped a car-load of fat hogs to Saint Joe, Missouri. Since he got two free railroad passes, he took me along with him. Up until that time, I had never been in a town of more than four thousand. When I landed in Saint Joe-a city of sixty thousand—I was agog with excitement. I saw skyscrapers six storeys high and—wonder of wonders—I saw a street-car. I can close my eyes now and still see and hear that street-car. After the most thrilling and exciting day of my life, Father and I took a train back to Ravenwood, Missouri. Arriving there at two o'clock in the morning, we had to walk four miles home to the farm. And here is the point of the story: I was so exhausted that I slept and dreamed as I walked. I have often slept while riding horseback. And I am alive to tell it!

When men are completely exhausted they sleep right through the thunder and horror and danger of war. Dr. Foster Kennedy, the famous neurologist, tells me that during the retreat of the Fifth British Army in 1918, he saw soldiers so exhausted that they fell on the ground where they were and fell into a sleep as sound as a coma. They didn't even wake up when he raised their eyelids with his fingers. And he says he noticed that invariably the pupils of the eyes were rolled upward in the sockets. "After that," says Dr. Kennedy, "when I had trouble sleeping, I would practice rolling up my eyeballs into this position, and I found that in a few seconds I would begin to yawn and feel sleepy. It was an automatic reflex over which I had no control."

No man ever committed suicide by refusing to sleep and no one ever will. Nature would force a man to sleep in spite of all his will power. Nature will let us go without food or water far longer than she will let us go without sleep.

Speaking of suicide reminds me of a case that Dr. Henry C. Link describes in his book, *The Rediscovery of Man*. Dr. Link is vice-president of The Psychological Corporation and he interviews many people who are worried and depressed. In his chapter "On Overcoming Fears and Worries", he tells about a patient who wanted to commit suicide. Dr. Link knew arguing would only make the matter worse, so he said to this man: "If you are going to commit suicide anyway, you might at least do it in a heroic fashion. Run around the block until you drop dead."

He tried it, not once but several times, and each time felt better, in his mind if not in his muscles. By the third night he had achieved what Dr. Link intended in the first place-he was so physically tired (and physically relaxed) that he slept like a log. Later he joined an athletic club and began to compete in competitive sports. Soon he was feeling so good he wanted to live forever!

So, to keep from worrying about insomnia, here are five rules:

1. If yon can't sleep, do what Samuel Untermyer did. Get up and work or read until you do feel sleepy.
2. Remember that no one was ever killed by lack of sleep. Worrying about insomnia usually causes far more damage than sleeplessness.
3. Try prayer-or repeat Psalm XXIII, as Jeanette MacDonald does.
4. Relax your body. Read the book *Release from Nervous Tension*.
5. Exercise. Get yourself so physically tired you can't stay awake.

Part Seven In A Nutshell - Six Ways To Prevent Fatigue And Worry And Keep Your Energy And Spirits High

RULE 1 Rest before you get tired.

RULE 2 Learn to relax at your work.

RULE 3 If you are a housewife, protect your health and appearance by relaxing at home.

RULE 4 Apply these four good working habits

a. Clear your desk of all papers except those relating to the immediate problem at hand.

b. Do things in the order of their importance.

c. When you face a problem, solve it then and there if you have the facts necessary to make a decision.

d. Learn to organise, deputise, and supervise.

RULE 5 To prevent worry and fatigue, put enthusiasm into your work.

RULE 5 Remember, no one was ever killed by lack of sleep. It is worrying about insomnia that does the damage-not the insomnia

✦ ✦ ✦

PART EIGHT

How To Find The Kind Of Work In Which You May Be Happy And Successful

Chapter 29

The Major Decision Of Your Life

(This chapter is addressed to young men and women who haven't yet found the work they want to do. If you are in that category, reading this chapter may have a profound effect upon the remainder of your life.)

If you are under eighteen, you will probably soon be called upon to make the two most important decisions of your life-decisions that will profoundly alter all the days of your years: decisions that may have far-reaching effects upon your happiness, your income, your health; decisions that may make or break you.

What are these two tremendous decisions?

First: How are you going to make a living? Are you going to be a farmer, a mail carrier, a chemist, a forest ranger, a stenographer, a horse dealer, a college professor, or are you going to run a hamburger stand ?

Second: Whom are you going to select to be the father or mother of your children?

Both of those great decisions are frequently gambles. "Every boy," says Harry Emerson Fosdick in his book, *The Power to See It Through*, "every boy is a gambler when he chooses a vocation. He must stake his life on it."

How can you reduce the gamble in selecting a vocation? Read on; we will tell you as best we can.

First, try, if possible, to find work that you enjoy. I once asked David M. Goodrich, Chairman of the Board, B. F. Goodrich Company-tyre manufacturers, what he considered the first requisite of success in business, and he replied: "Having a good time at your work. If you enjoy what you are doing," he said, "you may work long hours, but it won't seem like work at all. It will seem like play."

Edison was a good example of that. Edison—the unschooled newsboy who grew up to transform the industrial life of America—Edison, the man who often ate and slept in his laboratory and toiled there for eighteen hours a day. But it wasn't toil to him. "I never did a day's work in my life," he exclaimed. "It was all fun."

No wonder he succeeded!

I once heard Charles Schwab say much the same thing. He said: "A man can succeed at almost anything for which he has unlimited enthusiasm."

But how can you have enthusiasm for a job when you haven't the foggiest idea of what you want to do? "The greatest tragedy I know of," said Mrs. Edna Kerr, who once hired thousands of employees for the Dupont Company, and is now assistant director of industrial relations for the American Home Products Company, "The greatest tragedy I know of," she told me, "is that so many young people never discover what they really want to do. I think no one else is so much to be pitied as the person who gets nothing at all out of his work but his pay." Mrs. Kerr reports that even college graduates come to her and say: "I have a B.A. degree from Dartmouth [or an M.A. from Cornell]. Have you some kind of work I can do for your firm?" They don't know themselves what they are able to do, or even what they would like to do. Is it any wonder that so many men and women who start out in life with competent minds and rosy dreams end up at forty in utter frustration and even with a nervous breakdown? In fact, finding the right occupation is important even for your health. When Dr. Raymond Pearl, of Johns Hopkins, made a study, together with some insurance companies, to discover the factors that make for a long life, he placed "the right occupation" high on the list. He might have said, with Thomas Carlyle: "Blessed is the man who has found his work. Let him ask no other blessedness."

I recently spent an evening with Paul W. Boynton, employment supervisor for the Socony-Vacuum Oil Company. During the last twenty years he has interviewed more than seventy-five thousand people looking for jobs, and he has written the book *6 Ways to Get a Job*. I asked him: "What is the greatest mistake young people make today in looking for work?" "They don't know what they want to do," he said. "It is perfectly appalling to realise that a man will give more thought to buying a suit of clothes that will wear out in a few years than he will give to choosing the career on which his whole future depends-on which his whole future happiness and peace of mind are based!"

And so what? What can you do about it? You can take advantage of a new profession called vocational guidance. It may help you or harm you depending on the ability and character of the counselor you consult. This new profession isn't even within gunshot of perfection yet. It hasn't even reached the Model T stage. But it has a great future. How can you make use of this science? By finding out where, in your community, you can get vocational tests and vocational advice.

Such advice can only take the form of suggestions. You have to make the decisions. Remember that these counselors are far from infallible. They don't always agree with one another. They sometimes make ridiculous mistakes. For example, a vocational-guidance counselor advised one of my students to become a writer solely because she had a large vocabulary. How absurd! It isn't as simple as that. Good writing is the kind that transfers your thoughts and emotions to the reader—and to do that, you don't need a large vocabulary, but you do need ideas, experience, convictions, examples and excitement. The vocational counselor who advised this girl with a large vocabulary to become an author succeeded in doing only one thing: he turned an erstwhile happy stenographer into a frustrated, would-be novelist.

The point I am trying to make is that vocational-guidance experts, even as you and I, are not infallible. Perhaps you had better consult several of them and then interpret their findings in the sunlight of common sense.

You may think it strange that I am including a chapter like this in a book devoted to worry. But it isn't strange at all, when you understand how many of our worries, regrets, and frustrations are spawned by work we despise. Ask your father about it-or your neighbour or your boss. No less an intellectual giant than John Stuart Mill declared that industrial misfits are "among the heaviest losses of society". Yes, and among the unhappiest people on this earth are those same "industrial misfits" who hate their daily work!

Do you know the kind of man who "cracked up" in the Army? The man who was misplaced! I'm not talking about battle casualties, but about the men who cracked up in ordinary service. Dr. William Menninger, one of our greatest living psychiatrists, was in charge of the Army's neuro-psychiatric division during the war, and he says: "We learned much in the Army as to the importance of selection and of placement, of putting the right man in the right job. ... A conviction of the importance of the job at hand was extremely important. Where a man had no interest, where he felt he was misplaced, where he thought he was not appreciated, where he believed his talents were being misused, invariably we found a potential if not an actual psychiatric casualty."

Yes-and for the same reasons, a man may "crack up" in industry. If he despises his business, he can crack it up, too.

Take, for example, the case of Phil Johnson. Phil Johnson's father owned a laundry, so he gave his son a job, hoping the boy would work into the business. But Phil hated the laundry, so he dawdled, loafed, did what he had to do and not a lick more. Some days he was "absent". His father was so hurt to think he had a shiftless, ambitionless son that he was actually ashamed before his employees.

One day Phil Johnson told his father he wanted to be a mechanic-work in a machine shop. What? Go back to overalls? The old man was shocked. But Phil had his way. He worked in greasy dungarees. He did much harder work than was required at the laundry. He worked longer hours, and he whistled at his job! He took up engineering, learned about engines, puttered with machines-and when Philip Johnson died, in 1944, he was president of the Boeing Aircraft Company, and was making the

Flying Fortresses that helped to win the war! If he had stuck with the laundry, what would have happened to him and the laundry-especially after his father's death? My guess is he would have ruined the business, cracked it up and run it into the ground.

Even at the risk of starting family rows, I would like to say to young people: Don't feel compelled to enter a business or trade just because your family wants you to do it! Don't enter a career unless you want to do it! However, consider carefully the advice of your parents. They have probably lived twice as long as you have. They have gained the kind of wisdom that comes only from much experience and the passing of many years. But, in the last analysis, you are the one who has to make the final decision. You are the one who is going to be either happy or miserable at your work.

Now, having said this, let me give you the following suggestions-some of them warnings-about choosing your work:

1. Read and study the following five suggestions about selecting a vocational-guidance counselor. These suggestions are right from the horse's mouth. They were made by one of America's leading vocational-guidance experts, Professor Harry Dexter Kitson of Columbia University.

a. "Don't go to anyone who tells you that he has a magic system that will indicate your 'vocational aptitude'. In this group are phrenologists, astrologers, 'character analysts', handwriting experts. Their 'systems' do not work."

b. "Don't go to anyone who tells you that he can give you a test that will indicate what occupation you should choose. Such a person violates the principle that a vocational counselor must take into account the physical, social, and economic conditions surrounding the counselee; and he should render his service in the light of the occupational opportunities open to the counselee."

c. "Seek a vocational counselor who has an adequate library of information about occupations and uses it in the counseling process."

d. "A thorough vocational-guidance service generally requires more than one interview."
e. "Never accept vocational guidance by mail."

2. Keep out of business and professions that are already jam-packed and overflowing! There are many thousands of different ways of making a living. But do young people know this? Not unless they hire a swami to gaze into a crystal ball. The result? In one school, two-thirds of the boys confined their choices to five occupations-five out of twenty thousand-and four-fifths of the girls did the same. Small wonder that a few business and professions are overcrowded-small wonder that insecurity, worry, and "anxiety neuroses" are rampant at times among the white-collar fraternity! Beware of trying to elbow your way into such overcrowded fields as law, journalism, radio, motion pictures, and the "glamour occupations".

3. Stay out of activities where the chances are only one out of ten of your being able to make a living. As an example, take selling life insurance. Each year countless thousands of men-frequently unemployed men-start out trying to sell life insurance without bothering to find out in advance what is likely to happen to them! Here is approximately what does happen, according to Franklin L. Bettger, Real Estate Trust Building, Philadelphia. For twenty years Mr. Bettger was one of the outstandingly successful insurance salesmen in America. He declares that ninety per cent of the men who start selling life insurance get so heart-sick and discouraged that they give it up within a year. Out of the ten who remain, one man will sell ninety per cent of the insurance sold by the group of ten; and the other nine will sell only ten per cent. To put it another way: if you start selling life insurance, the chances are nine to one that you will fail and quit within twelve months, and the chances are only one in a hundred that you will make ten thousand a year out of it. Even if you remain at it, the chances are only one out of ten that you will be able to do anything more than barely scratch out a living.

4. Spend weeks-even months, if necessary-finding out all you can about an occupation before deciding to devote your life to

it! How? By interviewing men and women who have already spent ten, twenty, or forty years in that occupation.

These interviews may have a profound effect on your future. I know that from my own experience. When I was in my early twenties, I sought the vocational advice of two older men. As I look back now, I can see that those two interviews were turning points in my career. In fact, it would be difficult for me even to imagine what my life would have been like had I not had those two interviews.

How can you get these vocational-guidance interviews? To illustrate, let's suppose that you are thinking about studying to be an architect. Before you make your decision, you ought to spend weeks interviewing the architects in your city and in adjoining cities. You can get their names and addresses out of a classified telephone directory. You can call at their offices either with or without an appointment. If you wish to make an appointment, write them something like this:

"Won't you please do me a little favour? I want your advice. I am eighteen years old, and I am thinking about studying to be an architect. Before I make up my mind, I would like to ask your advice.

"If you are too busy to see me at your office, I would be most grateful if you would grant me the privilege of seeing you for half an hour at your home."

Here is a list of questions you could ask:

a. If you had your life to live over, would you become an architect again?
b. After you have sized me up, I want to ask you whether you think I have what it takes to succeed as an architect.
c. Is the profession of architecture overcrowded?
d. If I studied architecture for four years, would it be difficult for me to get a job? What kind of job would I have to take at first?
e. If I had average ability, how much could I hope to earn during the first five years?

f. What are the advantages and disadvantages of being an architect?
g. If I were your son, would you advise me to become an architect?

If you are timid, and hesitate to face a "big shot" alone, here are two suggestions that will help.

First, get a lad of your own age to go with you. The two of you will bolster up one another's confidence. If you haven't someone of your own age to go with you, ask your father to go with you.

Second, remember that by asking his advice you are paying this man a compliment. He may feel flattered by your request. Remember that adults like to give advice to young men and women. The architect will probably enjoy the interview.

If you hesitate to write letters asking for an appointment, then go to a man's office without an appointment and tell him you would be most grateful if he would give you a bit of advice.

Suppose you call on five architects and they are all too busy to see you (which isn't likely), call on five more. Some of them will see you and give you priceless advice—advice that may save you years of lost time and heartbreak.

Remember that you are making one of the two most vital and far-reaching decisions of your life. So, take time to get the facts before you act. If you don't, you may spend half a lifetime regretting it.

If you can afford to do so, offer to pay a man for a half-hour of his time and advice.

5. Get over the mistaken belief that you are fitted for only a single occupation! Every normal person can succeed at a number of occupations, and every normal person would probably fail in many occupations. Take myself, for example: if I had studied and prepared myself for the following occupations, I believe I would have had a good chance of achieving some small measure of success and also of enjoying my work. I refer to such occupations as farming, fruit growing, scientific agriculture, medicine, selling,

advertising, editing a country newspaper, teaching, and forestry. On the other hand, I am sure I would have been unhappy, and a failure, at bookkeeping, accounting, engineering, operating a hotel or a factory, architecture, all mechanical trades, and hundreds of other activities.

✦ ✦ ✦

PART NINE

How To Lessen Your Financial Worries

Chapter 30

"Seventy Per Cent Of All Our Worries ..."

If I knew how to solve everybody's financial worries, I wouldn't be writing this book, I would be sitting in the White House-right beside the President. But here is one thing I can do: I can quote some authorities on this subject and make some highly practical suggestions and point out where you can obtain books and pamphlets that will give you additional guidance.

Seventy per cent of all our worries, according to a survey made by the Ladies' Home Journal, are about money. George Gallup, of the Gallup Poll, says that his research indicates that most people believe that they would have no more financial worries if they could increase their income by only ten per cent. That is true in many cases, but in a surprisingly large number of cases it is not true. For example, while writing this chapter, I interviewed an expert on budgets: Mrs. Elsie Stapleton-a woman who spent years as financial adviser to the customers and employees of Wanamaker's Department Store in New York and of Gimbel's. She has spent additional years as an individual consultant, trying to help people who were frantic with worry about money. She has helped people in all kinds of income brackets, all the way from a porter who earned less than a thousand dollars a year to an executive earning one hundred thousand dollars a year. And this is what she told me: "More money is not the answer to most people's financial worries. In fact, I have often seen it happen that an increase in income accomplished nothing but an increase in spending-and an increase in headaches. What causes most people to worry," she said, "is not that they haven't enough money, but that they don't know how to spend the money they have!" ... [You snorted at that last sentence, didn't you? Well, before you snort again, please remember that Mrs. Stapleton did not say that was true of all people. She said: "most people". She didn't mean you. She meant your sisters and your cousins, whom you reckon by the dozens.]

A lot of readers are going to say: "I wish this guy Carnegie had my bills to meet, my obligations to keep up-on my weekly salary. If he did, I'll bet he would change his tune." Well, I have had my financial troubles: I have worked ten hours a day at hard physical labour in the cornfields and hay barns of Missouri-worked until my one supreme wish was to be free from the aching pains of utter physical exhaustion. I was paid for that grueling work not a dollar an hour, nor fifty cents, nor even ten cents. I was paid five cents an hour for a ten-hour day.

I know what it means to live for twenty years in houses without a bathroom or running water. I know what it means to sleep in bedrooms where the temperature is fifteen degrees below zero. I know what it means to walk miles to save a nickel car-fare and have holes in the bottom of my shoes and patches on the seat of my pants. I know what it means to order the cheapest dish on a restaurant menu, and to sleep with my trousers under the mattress because I couldn't afford to have them pressed by a tailor.

Yet, even during those times, I usually managed to save a few dimes and quarters out of my income because I was afraid not to. As a result of this experience, I realised that if you and I long to avoid debt and financial worries, then we have to do what a business firm does: we have to have a plan for spending our money and spend according to that plan. But most of us don't do that. For example, my good friend, Leon Shimkin, general manager of the firm that publishes this book, pointed out to me a curious blindness that many people have in regard to their money. He told me about a book-keeper he knows, a man who is a wizard at figures when working for his firm-yet when it comes to handling his personal finances! ... Well, if this man gets paid on Friday noon, let us say, he will walk down the street, see an overcoat in a store window that strikes his fancy, and buy it-never giving a thought to the fact that rent, electric lights, and all kinds of "fixed" charges have to come out of that pay envelope sooner or later. No-he has the cash in his pocket, and that's all that counts. Yet this man knows that if the company he works for conducted its business in such a slap-happy manner, it would end up in bankruptcy.

Here's something to consider-where your money is concerned, you're in business for yourself! And it is literally "your business" what you do with your money.

But what are the principles of managing our money? How do we begin to make a budget and a plan? Here are eleven rules.

RULE 1 **Get the facts down on paper.**

When Arnold Bennett started out in London fifty years ago to be a novelist, he was poor and hardpressed. So he kept a record of what he did with every six pence. Did he wonder where his money was going? No. He knew. He liked the idea so much that he continued to keep such a record even after he became rich, world-famous, and had a private yacht.

John D. Rockefeller, Sr., also kept a ledger. He knew to the penny just where he stood before he said his prayers at night and climbed into bed.

You and I, too, will have to get notebooks and start keeping records. For the rest of our lives? No, not necessarily. Experts on budgets recommend that we keep an accurate account of every nickel we spend for at least the first month-and, if possible, for three months. This is to give us an accurate record of where our money goes, so we can draw up a budget.

Oh, you know where your money goes? Well, maybe so; but if you do, you are one in a thousand! Mrs. Stapleton tells me it is a common occurrence for men and women to spend hours giving her facts and figures, so she can get them down on paper-then, when they see the result on paper, they exclaim: "Is that the way my money goes?" They can hardly believe it. Are you like that? Could be.

RULE 2 Get a tailor-made budget that really fits your needs.

Mrs. Stapleton tells me that two families may live side by side in identical houses, in the very same suburb, have the same number of children in the family, and receive the same income-yet their budgeting needs will be radically different. Why? Because people are different. She says a budget has to be a personal, custom-made job.

The idea of a budget is not to wring all the joy out of life. The idea is to give us a sense of material security-which in many cases means emotional security and freedom from worry. "People who live on budgets," Mrs. Stapleton told me, "are happier people."

But how do you go about it? First, as I said, you must list all expenses. Then get advice. In many cities of twenty thousand and up, you will find family-welfare societies that will gladly give you free advice on financial problems and help you draw up a budget to fit your income.

RULE 3 Learn how to spend wisely.

By this I mean: learn how to get the best value for your money. All large corporations have professional buyers and purchasing agents who do nothing but get the very best buys for their firms. As steward and manager of your personal estate, why shouldn't you do likewise?

RULE 4 Don't increase your headaches with your income.

Mrs. Stapleton told me that the budgets she dreads most to be called into consultation on are family incomes of five thousand dollars a year. I asked her why. "Because," she said, "five thousand a year seems to be a goal to most American families. They may go along sensibly and sanely for years—then, when their income rises to five thousand a year, they think they have 'arrived'. They start branching out. Buy a house in the suburbs, 'that doesn't cost any more than renting an apartment'. Buy a car, a lot of new furniture, and a lot of new clothes—and the first thing you know, they are running into the red. They are actually less happy than they were the first before, because they have bitten off too much with their increase in income."

That is only natural. We all want to get more out of life. But in the long run, which is going to bring us more happiness-forcing ourselves to live within a tight budget, or having dunning letters in the mail and creditors pounding on the front door?

RULE 5 Try to build credit, in the event you must borrow.

If you are faced with an emergency and find you must borrow, life-insurance policies, Defence Bonds and Savings Certificates are literally money in your pocket. However, be sure your insurance policies have a savings aspect, if you want to borrow on them, for this means a cash value. Certain types of insurance, called "term insurance", are merely for your protection over a given period of time and do not build up reserves. These policies are obviously of no use to you for borrowing purposes. Therefore, the rule is: Ask questions! Before you sign for a policy, find out if it has a cash value in case you have to raise money.

Now, suppose you haven't insurance you can borrow on, and you haven't any bonds, but you do own a house, or a car, or some other kind of collateral. Where do you go to borrow? By all means, to a bank! Banks all over this land are subject to strict regulation; they have a reputation to maintain in the community; the rate of interest they can charge is fixed firmly by law; and they will deal with you fairly. Frequently, if you are in a financial jam, the bank will go so far as to discuss your problems with you, make a plan, and help you work your way out of your worry and indebtedness. I repeat, I repeat, if you have collateral, go to a bank!

However, suppose you are one of the thousands who don't have collateral, don't own any property, and have nothing to offer as guarantee except your wages or salary? Then, as you value your life, heed this word of warning! Do not-do not-apply to the first "loan company" whose alluring advertisements you see in the paper. These people, to read some of their ads, are as generous as Santa Claus. Don't you believe it! However, there are some companies that are ethical, honest, and strictly on the level. They are doing a service to those people who are faced with illness or emergency and have to raise money. They charge a higher rate of interest than the banks, but they have to do this, for they take greater risks and have greater expenses in collecting. But, before doing business with any loan company, go to your bank, talk to one of its officers, and ask him to recommend a loan company that he knows to be fair. Otherwise-otherwise-well, I don't want to give you nightmares, but here is what can happen:

At one time a newspaper in Minneapolis conducted an investigation into loan companies that were supposedly operating within the regulations laid down by the Russell Sage Foundation. I know a man who worked on that investigation-his name is Douglas Lurton, and he is now editor of Your Life magazine. Doug Lurton tells me that the abuses he saw among the poorer class of debtors would make your hair stand on end. Loans that had begun as a mere fifty dollars had soared and multiplied to three and four hundred dollars before they were paid. Wages were garnished; and, frequently, the man whose wages were attached was fired by his company. In numerous instances, when the man was unable to pay, the loan sharks simply sent an appraiser into his home to "evaluate" his furniture-and cleaned out the home! People were found who had been paying on small loans for four and five years and still owed money! Unusual cases? To quote Doug Lurton: "In our campaign, we so flooded the court with cases of this sort that the judges cried uncle, and the newspaper itself had to set up an arbitration bureau to take care of the hundreds of cases."

How is such a thing possible? Well, the answer, of course, is in all sorts of hidden charges and extra "legal fees". Here is a rule to remember in dealing with loan companies: if you are absolutely certain, beyond the shadow of a doubt, that you can pay the money off quickly, then your interest will be low, or reasonably low, and you will get off fairly. But if you have to renew, and keep on renewing, then your interest can mount into figures that would make Einstein dizzy. Doug Lurton tells me that in some cases these additional fees had swollen the original indebtedness to two thousand per cent, or about five hundred times as much as a bank would charge!

RULE 6 **Protect yourself against illness, fire, and emergency expenses.**

Insurance is available, for relatively small sums, on all kinds of accidents, misfortunes, and conceivable emergencies. I am not suggesting that you cover yourself for everything from slipping in the bathtub to catching German measles-but I do suggest that you protect yourself against the major misfortunes that you know could cost you money and therefore do cost you worry. It's cheap at the price.

For example, I know a woman who had to spend ten days in a hospital last year and, when she came out, was presented a bill-for exactly eight dollars! The answer? She had hospital insurance.

RULE 7 **Do not have your life-insurance proceeds paid to your widow in cash.**

If you are carrying life insurance to provide for your family after you're gone, do not, I beg of you, have your insurance paid in one lump sum.

What happens to "a new widow with new money"? I'll let Mrs. Marion S. Eberly answer that question. She is head of the Women's Division of the Institute of Life Insurance, 60 East 42nd Street, New York City. She speaks before women's clubs all over America on the wisdom of using life-insurance proceeds to purchase a life income for the widow instead of giving her the proceeds in cash. She tells me one widow who received twenty thousand dollars in cash and lent it to her son to start in the autoaccessory business. The business failed, and she is destitute now. She tells of another widow who was persuaded by a slick real-estate salesman to put most of her life-insurance money in vacant lots that were "sure to double in value within a year". Three years later, she sold the lots for one-tenth of what she paid for them. She tells of another widow who had to apply to the Child Welfare Association for the support of her children-within twelve months after she had been left fifteen thousand dollars in life insurance. A hundred thousand similar tragedies could be told.

"The average lifetime of twenty-five thousand dollars left in the hands of a woman is less than seven years." That statement was made by Sylvia S. Porter, financial editor of the New York Post, in the Ladies' Home Journal.

Years ago, The Saturday Evening Post said in an editorial: "The ease with which the average widow without business training, and with no banker to advise her, can be wheedled into putting her husband's life-insurance money into wildcat stocks by the first slick salesman who approaches her is proverbial. Any lawyer or banker can cite a dozen cases in which the entire savings of a thrifty man's lifetime, amassed by years of sacrifice and self-denial, were swept away simply because a widow or an orphan trusted one of the slick crooks who rob women for a livelihood."

If you want to protect your widow and your children, why not take a tip from J. P. Morgan-one of the wisest financiers who ever lived. He left money in his will to sixteen principal legatees. Twelve were women. Did he leave these women cash? No. He left trust funds that ensured these women a monthly income for life.

RULE 8 **Teach your children a responsible attitude toward money.**

I shall never forget an idea I once read in Your Life magazine. The author, Stella Weston Turtle, described how she was teaching her little girl a sense of responsibility about money. She got an extra cheque-book from the bank and gave it to her nine-year-old daughter. When the daughter was given her weekly allowance, she "deposited" the money with her mother, who served as a bank for the child's funds. Then, throughout the week, whenever she wanted a cent or two, she "drew a cheque" for that amount and kept track of her balance. The little girl not only found that fun, but began to learn real responsibility in handling her money.

This is an excellent method and if you have a son or daughter of school age, and you want this child to learn how to handle money, I recommend it for your consideration.

RULE 9 If necessary, make a little extra money off your kitchen stove.

If after you budget your expenses wisely you still find that you don't have enough to make ends meet, you can then do one of two things: you can either scold, fret, worry, and complain, or you can plan to make a little additional money on the side. How? Well, all you have to do to make money is to fill an urgent need that isn't being adequately filled now. That is what Mrs. Nellie Speer, 37-09 83rd Street, Jackson Heights, New York, did. In 1932, she found herself living alone in a three-room apartment. Her husband had died, and both of her children were married. One day, while having some ice-cream at a drug-store soda fountain, she noticed that the fountain was also selling bakery pies that looked sad and dreary. She asked the proprietor if he would buy some real home-made pies from her. He ordered two. "Although I was a good cook," Mrs. Speer said, as she told me the story, "I had always had servants when we lived in Georgia, and I had never baked more than a dozen pies in my life. After getting that order for two pies, I asked a neighbour woman how to cook an apple-pie. The sodafountain customers were delighted with my first two home-baked pies, one apple, one lemon. The drugstore ordered five the next day. Then orders gradually came in from other fountains and luncheonettes. Within two years, I was baking five thousand pies a year-I was doing all the work myself in my own tiny kitchen, and I was making a thousand dollars a year clear, without a penny's expense except the ingredients that went into the pies."

The demand for Mrs. Speer's home-baked pastry became so great that she had to move out of her kitchen into a shop and hire two girls to bake for her: pies, cakes, bread, and rolls. During the war, people stood in line for an hour at a time to buy her home-baked foods.

"I have never been happier in my life," Mrs. Speer said. "I work in the shop twelve to fourteen hours a day, but I don't get tired because it isn't work to me. It is an adventure in living. I am doing my part to make people a little happier. I am too busy to be lonesome or worried. My work has filled a gap in my life left vacant by the passing of my mother and husband and my home."

When I asked Mrs. Speer if she felt that other women who were good cooks could make money in their spare time in a similar way, in towns of ten thousand and up, she replied: "Yes-of course they can!"

Mrs. Ora Snyder will tell you the same thing. She lives in a town of thirty thousand-Maywood, Illinois. Yet she started in business with the kitchen stove and ten cents' worth of ingredients. Her husband fell ill. She had to earn money. But how? No experience. No skill. No capital. Just a housewife. She took the white of an egg and sugar and made some candy on the back of the kitchen stove; then she took her pan of candy and stood near the school and sold it to the children for a penny a piece as they went home. "Bring more pennies tomorrow," she said. "I'll be here every day with my home-made candy." During the first week, she not only made a profit, but had also put a new zest into living. She was making both herself and the children happy. No time now for worry.

This quiet little housewife from Maywood, Illinois, was so ambitious that she decided to branch out—to have an agent sell her kitchen-made candy in roaring, thundering Chicago. She timidly approached an Italian selling peanuts on the street. He shrugged his shoulders. His customers wanted peanuts, not candy. She gave him a sample. He liked it, began selling her candy, and made a good profit for Mrs. Snyder on the first day.

Four years later, she opened her first store in Chicago. It was only eight feet wide. She made her candy at night and sold it in the daytime. This erstwhile timid housewife, who started her candy factory on her kitchen stove, now has seventeen stores-fifteen of them in the busy Loop district of Chicago.

Here is the point I am trying to make. Nellie Speer, in Jackson Heights, New York, and Mrs. Ora Snyder, in Maywood, Illinois, instead of worrying about finances, did something positive. They started in an extremely small way to make money off the kitchen stove-no overhead, no rent, no advertising, no salaries. Under these conditions, it is almost impossible for a woman to be defeated by financial worries.

Look around you. You will find many needs that are not filled. For example, if you train yourself to be a good cook, you can probably make money by starting cooking classes for young girls right in your own kitchen. You can get your students by ringing door-bells.

Books have been written about how to make money in your spare time; inquire at your public library. There are many opportunities for both men and women. But one word of warning: unless you have a natural gift for selling, don't attempt door-to-door selling. Most people hate it and fail at it.

RULE 10 Don't gamble-ever.

I am always astounded by the people who hope to make money by betting on the ponies or playing slot machines. I know a man who makes his living by owning a string of these "one armed bandits", and he has nothing but contempt for the foolish people who are so naive as to imagine that they can beat a machine that is already rigged against them.

I also know one of the best known bookmakers in America. He was a student in my adult-education classes. He told me that with all his knowledge of horse racing, he couldn't make money betting on the ponies. Yet the facts are that foolish people bet six billion dollars a year on the races—six times as much as our total national debt back in 1910. This bookmaker also told me that if he had an enemy he despised, he could think of no better way of ruining him than by getting him to bet on the races. When I asked him what would happen to the man who played the races according to the tipster sheets, he replied: "You could lose the Mint by betting that way."

If we are determined to gamble, let's at least be smart. Let's find out what the odds are against us. How? By reading a book entitled *How to Figure the Odds*, by Oswald Jacoby-an authority on bridge and poker, a top-ranking mathematician, a professional statistician, and an insurance actuary. This book devotes 215 pages to telling you what the odds are against your winning when you play the ponies, roulette, craps, slot machines, draw poker, stud poker, contract bridge, auction pinochle, the stock market. This book also give you the scientific, mathematical chances on a score of other activities. It doesn't pretend to show how to make money gambling. The author has no axe to grind. He merely shows you what the odds are against your winning in all the usual ways of gambling; and when you see the odds, you will pity the poor suckers who stake their hard-earned wages on horse races or cards or dice or slot machines. If you are tempted to shoot craps or play poker or bet on horses, this book may save you a hundred times-yes, maybe a thousand times-what it costs.

RULE 11 **If we can't possibly improve our financial situation, let's be good to ourselves and stop resenting what can't be changed.**

If we can't possibly improve our financial situation, maybe we can improve our mental attitude towards it. Let's remember that other people have their financial worries, too. We may be worried because we can't keep up with the Joneses; but the Joneses are probably worried because they can't keep up with the Ritzes; and the Ritzes are worried because they can't keep up with the Vanderbilts.

Some of the most famous men in American history have had their financial troubles. Both Lincoln and Washington had to borrow money to make the trip to be inaugurated as President.

If we can't have all we want, let's not poison our days and sour our dispositions with worry and resentment. Let's be good to ourselves. Let's try to be philosophical about it. "If you have what seems to you insufficient," said one of Rome's greatest philosophers, Seneca, "then you will be miserable even if you possess the world."

And let's remember this: even if we owned the entire United States with a hog-tight fence around it, we could eat only three meals a day and sleep in only one bed at a time.

To lessen financial worries, let's try to follow these eleven rules:

1 **GOLDEN RULE 1**

Get the facts down on paper.

2 **GOLDEN RULE 2**

Get a tailor-made budget that really fits your needs.

3 **GOLDEN RULE 3**

Learn how to spend wisely.

4 **GOLDEN RULE 4**

Don't increase your headaches with your income.

5 **GOLDEN RULE 5**

Try to build credit, in the event you must borrow.

6 **GOLDEN RULE 6**

Protect yourself against illness, fire, and emergency expenses.

7 GOLDEN RULE 7

Do not have your life-insurance proceeds paid to your widow in cash.

8 GOLDEN RULE 8

Teach your children a responsible attitude towards money.

9 GOLDEN RULE 9

If necessary, make a little extra money off your kitchen stove.

10 GOLDEN RULE 10

Don't gamble-ever.

11 GOLDEN RULE 11

If we can't possibly improve our financial situation, let's be good to ourselves and stop resenting what can't be changed.

PART TEN

"How I Conquered Worry" - 32 True Stories

Six Major Troubles Hit Me All At Once

By C.I. Black Wood
Proprietor, Blackwood-Davis Business College Oklahoma City, Oklahoma

In the summer of 1943, it seemed to me that half the worries of the world had come to rest on my shoulders.

For more than forty years, I had lived a normal, carefree life with only the usual troubles which come to a husband, father, and businessman. I could usually meet these troubles easily, but suddenly-wham! wham!! wham!!! wham! !!! WHAM! !!!! WHAM!!!!!! Six major troubles hit me all at once. I pitched and tossed and turned in bed all night long, half dreading to see the day come, because I faced these six major worries.

1. My business college was trembling on the verge of financial disaster because all the boys were going to war; and most of the girls were making more money working in war plants without training than my graduates could make in business offices with training.

2. My older son was in service, and I had the heart-numbing worry common to all parents whose sons were away at war.

3. Oklahoma City had already started proceedings to appropriate a large tract of land for an airport, and my home- formerly my father's home-was located in the centre of this tract. I knew that I would be paid only one tenth of its value, and, what was even worse, I would lose my home; and because of the housing shortage, I worried about whether I could possibly find another home to shelter my family of six. I feared we might have to live in a tent. I even worried about whether we would be able to buy a tent.

4. The water well on my property went dry because a drainage canal had been dug near my home. To dig a new well would be throwing five hundred dollars away because the land was probably being appropriated. I had to carry water to my livestock in buckets every morning for two months, and I feared I would have to continue it during the rest of the war.

5. I lived ten miles away from my business school and I had a class B petrol card: that meant I couldn't buy any new tyres, so I worried about how I could ever get to work when the superannuated tyres on my old Ford gave up the ghost.

6. My oldest daughter had graduated from high school a year ahead of schedule. She had her heart set on going to college, and I just didn't have the money to send her. I knew her heart would be broken.

One afternoon while sitting in my office, worrying about my worries, I decided to write them all down, for it seemed no one ever had more to worry about than I had. I didn't mind wrestling with worries that gave me a fighting chance to solve them, but these worries all seemed to be utterly beyond my control. I could do nothing to solve them. So I filed away this typewritten list of my troubles, and, as the months passed, I forgot that I had ever written it. Eighteen months later, while transferring my files, I happened to come across this list of my six major problems that had once threatened to wreck my health. I read them with a great deal of interest-and profit. I now saw that not one of them had come to pass.

Here is what had happened to them:

1. I saw that all my worries about having to close my business college had been useless because the government had started paying business schools for training veterans and my school was soon filled to capacity.

2. I saw that all my worries about my son in service had been useless: he was coming through the war without a scratch.

3. I saw that all my worries about my land being appropriated for use as an airport had been useless because oil had been struck within a mile of my farm and the cost for procuring the land for an airport had become prohibitive.

4. I saw that all my worries about having no well to water my stock had been useless because, as soon as I knew my land would not be appropriated, I spent the money necessary to dig a new well to a deeper level and found an unfailing supply of water.

5. I saw that all my worries about my tyres giving out had been useless, because by recapping and careful driving, the tyres had managed somehow to survive.

6. I saw that all my worries about my daughter's education had been useless, because just sixty days before the opening of college, I was offered-almost like a miracle-an auditing job which I could do outside of school hours, and this job made it possible for me to send her to college on schedule.

I had often heard people say that ninety-nine per cent of the things we worry and stew and fret about never happen, but this old saying didn't mean much to me until I ran across that list of worries I had typed out that dreary afternoon eighteen months previously.

I am thankful now that I had to wrestle in vain with those six terrible worries. That experience has taught me a lesson I'll never forget. It has shown me the folly and tragedy of stewing about events that haven't happened-events that are beyond our control and may never happen.

Remember, today is the tomorrow you worried about yesterday. Ask yourself: How do I KNOW this thing I am worrying about will really come to pass?

✦ ✦ ✦

I Can Turn Myself in to a Shouting Optimist Within an Hour

By Roger W. Babson
Famous Economist Babson Park, Wellesley Hills, Massachusetts

When I find myself depressed over present conditions, I can, within one hour, banish worry and turn myself into a shouting optimist.

Here is how I do it. I enter my library, close my eyes, and walk to certain shelves containing only books on history. With my eyes still shut, I reach for a book, not knowing whether I am picking up Prescott's *Conquest of Mexico* or Suetonius' *Lives of the Twelve Caesars*. With my eyes still closed, I open the book at random. I then open my eyes and read for an hour; and the more I read, the more sharply I realise that the world has always been in the throes of agony, that civilisation has always been tottering on the brink. The pages of history fairly shriek with tragic tales of war, famine, poverty, pestilence, and man's inhumanity to man. After reading history for an hour, I realise that bad as conditions are now, they are infinitely better than they used to be. This enables me to see and face my present troubles in their proper perspective as well as to realise that the world as a whole is constantly growing better.

Here is a method that deserves a whole chapter. Read history! Try to get the viewpoint of ten thousand years-and see how trivial your troubles are, in terms of eternity.

✦ ✦ ✦

How I Got Rid Of An Inferiority Complex

By Elmer Thomas
United States Senator from Oklahoma

When I was fifteen I was constantly tormented by worries and fears and self-consciousness. I was extremely tall for my age and as thin as a fence rail. I stood six feet two inches and weighed only 118 pounds. In spite of my height, I was weak and could never compete with the other boys in baseball or running games. They poked fun at me and called me "hatch-face". I was so worried and self-conscious that I dreaded to meet anyone, and I seldom did, for our farmhouse was off the public road and surrounded by thick virgin timber that had never been cut since the beginning of time. We lived half a mile from the highway; and a week would often go by without my seeing anyone except my mother, father, and brothers and sisters.

I would have been a failure in life if I had let those worries and fears whip me. Every day and every hour of the day, I brooded over my tall, gaunt, weak body. I could hardly think of anything else. My embarrassment, my fear, was so intense that it is almost impossible to describe it. My mother knew how I felt. She had been a school-teacher, so she said to me: "Son, you ought to get an education, you ought to make your living with your mind because your body will always be a handicap."

Since my parents were unable to send me to college, I knew I would have to make my own way; so I hunted and trapped opossum, skunk, mink, and raccoon one winter; sold my hides for four dollars in the spring, and then bought two little pigs with my four dollars. I fed the pigs slop and later corn and sold them for forty dollars the next fall. With the proceeds from the sale of the two hogs I went away to the Central Normal College-located at Danville, Indiana. I paid a dollar and forty cents a week for my board and fifty cents a week for my room. I wore a brown shirt my mother had made me. (Obviously, she used brown cloth because it wouldn't show the dirt.) I wore a suit of clothes that had once belonged to my father. Dad's clothes didn't fit me and neither did his old congress gaiter shoes that I wore-shoes that had elastic bands in the sides that stretched when you put them on. But the stretch had long since gone out of the bands, and the tops were so loose that the shoes almost dropped off my feet as I walked. I was embarrassed to associate with the other students, so I sat in my room alone and studied. The deepest desire of my life was to be able to buy some store clothes that fit me, clothes that I was not ashamed of.

Shortly after that, four events happened that helped me to overcome my worries and my feeling of inferiority. One of these events gave me courage and hope and confidence and completely changed all the rest of my life. I'll describe these events briefly:

First: After attending this normal school for only eight weeks, I took an examination and was given a third-grade certificate to teach in the country public schools. To be sure, this certificate was good for only six months, but it was fleeting evidence that somebody had faith in me-the first evidence of faith that I ever had from anyone except my mother.

Second: A country school board at a place called Happy Hollow hired me to teach at a salary of two dollars per day, or forty dollars per month. Here was even more evidence of somebody's faith in me.

Third: As soon as I got my first cheque I bought some store clothes-clothes that I wasn't ashamed to wear. If someone gave me a million dollars now, it wouldn't thrill me half as much as that first suit of store clothes for which I paid only a few dollars.

Fourth: The real turning point in my life, the first great victory in my struggle against embarrassment and inferiority occurred at the Putnam County Fair held annually in Bain-bridge, Indiana. My mother had urged me to enter a public-speaking contest that was to be held at the fair. To me, the very idea seemed fantastic. I didn't have the courage to talk even to one person-let alone a crowd. But my mother's faith in me was almost pathetic. She dreamed great dreams for my future. She was living her own life over in her son. Her faith inspired me to enter the contest. I chose for my subject about the last thing in the world that I was qualified to talk on: "The Fine and Liberal Arts of America". Frankly, when I began to prepare a speech I didn't know what the liberal arts were, but it didn't matter much because my audience didn't know, either.

I memorised my flowery talk and rehearsed it to the trees and cows a hundred times. I was so eager to make a good showing for my mother's sake that I must have spoken with emotion. At any rate, I was awarded the first prize. I was astounded at what happened. A cheer went up from the crowd. The very boys who had once ridiculed me and poked fun at me and called me hatchet-faced now slapped me on the back and said: "I knew you could do it, Elmer." My mother put her arms around me and sobbed. As I look back in retrospect, I can see that winning that speaking contest was the turning point of my life. The local newspapers ran an article about me on the front page and prophesied great things for my future. Winning that contest put me on the map locally and gave me prestige, and, what is far more important, it multiplied my confidence a hundredfold. I now realise that if I had not won that contest, I probably would never have become a member of the United States Senate, for it lifted my sights, widened my horizons, and made me realise that I had latent abilities that I never dreamed I possessed. Most important, however, was the fact that the first prize in the oratorical contest was a year's scholarship in the Central Normal College.

I hungered now for more education. So, during the next few years-from 1896 to 1900-I divided my time between teaching and studying. In order to pay my expenses at De Pauw University, I waited on tables, looked after furnaces, mowed lawns, kept books, worked in the wheat and cornfields during the summer, and hauled gravel on a public road-construction job.

In 1896, when I was only nineteen, I made twenty-eight speeches, urging people to vote for William Jennings Bryan for President. The excitement of speaking for Bryan aroused a desire in me to enter politics myself. So when I entered De Pauw University, I studied law and public speaking. In 1899 I represented the university in a debate with Butler College, held in Indianapolis, on the subject "Resolved that United States Senators should be elected by popular vote." I won other speaking contests and became editor-in-chief of the class of 1900 College Annual, The Mirage, and the university paper, The Palladium.

After receiving my A.B. degree at De Pauw, I took Horace Greeley's advice-only I didn't go west, I went south-west. I went down to a new country: Oklahoma. When the Kiowa, Comanche, and Apache Indian reservation was opened, I home-steaded a claim and opened a law office in Lawton, Oklahoma. I served in the Oklahoma State Senate for thirteen years, in the lower House of Congress for four years, and at fifty years of age, I achieved my lifelong ambition: I was elected to the United States Senate from Oklahoma. I have served in that capacity since March 4, 1927. Since Oklahoma and Indian Territories became the state of Oklahoma on November 16, 1907, I have been continuously honoured by the Democrats of my adopted state by nominations-first for State Senate, then for Congress, and later for the United States Senate.

I have told this story, not to brag about my own fleeting accomplishments, which can't possibly interest anyone else. I have told it wholly with the hope that it may give renewed courage and confidence to some poor boy who is now suffering from the worries and shyness and feeling of inferiority that devastated my life when I was wearing my father's cast-off clothes and gaiter shoes that almost dropped off my feet as I walked.

(Editor's note: It is interesting to know that Elmer Thomas, who was so ashamed of his ill-fitting clothes as a youth, was later voted the best-dressed man in the United States Senate.)

✦ ✦ ✦

I Lived In The Garden Of Allah

By R.V.C. Bodley
Descendant of Sir Thomas Bodley, founder of the Bodleian Library, Oxford Author of Wind in the Sahara, The Messenger, and fourteen other volumes

IN 1918, I turned my back on the world I had known and went to north-west Africa and lived with the Arabs in the Sahara, the Garden of Allah. I lived there seven years. I learned to speak the language of the nomads. I wore their clothes, I ate their food, and adopted their mode of life, which has changed very little during the last twenty centuries. I became an owner of sheep and slept on the ground in the Arabs' tents. I also made a detailed study of their religion. In fact, I later wrote a book about Mohammed, entitled *The Messenger*.

Those seven years which I spent with these wandering shepherds were the most peaceful and contented years of my life.

I had already had a rich and varied experience: I was born of English parents in Paris; and lived in France for nine years. Later I was educated at Eton and at the Royal Military College at Sandhurst. Then I spent six years as a British army officer in India, where I played polo, and hunted, and explored in the Himalayas as well as doing some soldiering. I fought through the First World War and, at its close, I was sent to the Paris Conference as an assistant military attaché. I was shocked and disappointed at what I saw there. During the four years of slaughter on the Western Front, I had believed we were fighting to save civilisation. But at the Paris Peace Conference, I saw selfish politicians laying the groundwork for the Second World War-each country grabbing all it could for itself, creating national antagonisms, and reviving the intrigues of secret diplomacy.

I was sick of war, sick of the army, sick of society. For the first time in my career, I spent sleepless nights, worrying about what

I should do with my life. Lloyd George urged me to go in for politics. I was considering taking his advice when a strange thing happened, a strange thing that shaped and determined my life for the next seven years. It all came from a conversation that lasted less than two hundred seconds-a conversation with "Ted" Lawrence, "Lawrence of Arabia", the most colourful and romantic figure produced by the First World War. He had lived in the desert with the Arabs and he advised me to do the same thing. At first, it sounded fantastic.

However, I was determined to leave the army, and I had to do something. Civilian employers did not want to hire men like me-ex-officers of the regular army-especially when the labour market was jammed with millions of unemployed. So I did as Lawrence suggested: I went to live with the Arabs. I am glad I did so. They taught me how to conquer worry. Like all faithful Moslems, they are fatalists. They believe that every word Mohammed wrote in the Koran is the divine revelation of Allah. So when the Koran says: "God created you and all your actions," they accept it literally. That is why they take life so calmly and never hurry or get into unnecessary tempers when things go wrong. They know that what is ordained is ordained; and no one but God can alter anything. However, that doesn't mean that in the face of disaster, they sit down and do nothing. To illustrate, let me tell you of a fierce, burning windstorm of the sirocco which I experienced when I was living in the Sahara. It howled and screamed for three days and nights. It was so strong, so fierce, that it blew sand from the Sahara hundreds of miles across the Mediterranean and sprinkled it over the Rhone Valley in France. The wind was so hot I felt as if the hair was being scorched off my head. My throat was parched. My eyes burned. My teeth were full of grit. I felt as if I were standing in front of a furnace in a glass factory. I was driven as near crazy as a man can be and retain his sanity. But the Arabs didn't complain. They shrugged their shoulders and said: "Mektoub!" ... "It is written."

But immediately after the storm was over, they sprang into action: they slaughtered all the lambs because they knew they would die anyway; and by slaughtering them at once, they hoped to save the mother sheep. After the lambs were slaughtered, the flocks were driven southward to water. This was all done calmly, without worry or complaining or mourning over their losses. The tribal chief said: "It is not too bad. We might have lost everything. But praise God, we have forty per cent of our sheep left to make a new start."

I remember another occasion, when we were motoring across the desert and a tyre blew out. The chauffeur had forgotten to mend the spare tyre. So there we were with only three tyres. I fussed and fumed and got excited and asked the Arabs what we were going to do. They reminded me that getting excited wouldn't help, that it only made one hotter. The blown-out tyre, they said, was the will of Allah and nothing could be done about it. So we started on, crawling along on the rim of a wheel. Presently the car spluttered and stopped. We are out of petrol. The chief merely remarked: "Mektoub!" and, there again, instead of shouting at the driver because he had not taken on enough petrol, everyone remained calm and we walked to our destination, singing as we went.

The seven years I spent with the Arabs convinced me that the neurotics, the insane, the drunks of America and Europe are the product of the hurried and harassed lives we live in our so-called civilisation.

As long as I lived in the Sahara, I had no worries. I found there, in the Garden of Allah, the serene contentment and physical well-being that so many of us are seeking with tenseness and despair.

Many people scoff at fatalism. Maybe they are right. Who knows? But all of us must be able to see how our fates are often determined for us. For example, if I had not spoken to Lawrence of Arabia at three minutes past noon on a hot August day in 1919, all the years that have elapsed since then would have been completely different. Looking back over my life, I can see how it has been shaped and moulded time and again by events far beyond my control. The Arabs call it mektoub, kismet-the will of Allah. Call it anything you wish. It does strange things to you. I only know that today-seventeen years after leaving the Sahara-I still maintain that happy resignation to the inevitable which I learned from the Arabs. That philosophy has done more to settle my nerves than a thousand sedatives could have achieved.

You and I are not Mohammedans: we don't want to be fatalists. But when the fierce, burning winds blow over our lives-and we cannot prevent them-let us, too, accept the inevitable. And then get busy and pick up the pieces.

⁂

Five Methods I Use To Banish Worry

By Professor William Lyon Phelps

[I had the privilege of spending an afternoon with Billy Phelps, of Yale, shortly before his death. Here are the five methods he used to banish worry-based on the notes I took during that interview. -DALE CARNEGIE]

1 When I was twenty-four years old, my eyes suddenly gave out. After reading three or four minutes, my eyes felt as if they were full of needles; and even when I was not reading, they were so sensitive that I could not face a window. I consulted the best occultists in New Haven and New York. Nothing seemed to help me. After four o'clock in the afternoon, I simply sat in a chair in the darkest corner of the room, waiting for bedtime. I was terrified. I feared that I would have to give up my career as a teacher and go out West and get a job as a lumberjack. Then a strange thing happened which shows the miraculous effects of the mind over physical ailments. When my eyes were at their worst that unhappy winter, I accepted an invitation to address a group of undergraduates.

The hall was illuminated by huge rings of gas jets suspended from the ceiling. The lights pained my eyes so intensely that, while sitting on the platform, I was compelled to look at the floor. Yet during my thirty-minute speech, I felt absolutely no pain, and I could look directly at these lights without any blinking whatever. Then when the assembly was over, my eyes pained me again.

I thought then that if I could keep my mind strongly concentrated on something, not for thirty minutes, but for a week, I might be cured. For clearly it was a case of mental excitement triumphing over a bodily illness.

I had a similar experience later while crossing the ocean. I had an attack of lumbago so severe that I could not walk. I suffered extreme pain when I tried to stand up straight. While in that condition, I was invited to give a lecture on shipboard. As soon as I began to speak, every trace of pain and stiffness left my body; I stood up straight, moved about with perfect flexibility, and spoke for an hour. When the lecture was over, I walked away to my stateroom with ease. For a moment, I thought I was cured. But the cure was only temporary. The lumbago resumed its attack.

These experiences demonstrated to me the vital importance of one's mental attitude. They taught me the importance of enjoying life while you may. So I live every day now as if it were the first day I had ever seen and the last I were going to see. I am excited about the daily adventure of living, and nobody in a state of excitement will be unduly troubled with worries. I love my daily work as a teacher. I wrote a book entitled *The Excitement of Teaching*. Teaching has always been more than an art or an occupation to me. It is a passion. I love to teach as a painter loves to paint or a singer loves to sing. Before I get out of bed in the morning, I think with ardent delight of my first group of students. I have always felt that one of the chief reasons for success in life is enthusiasm.

2 I have found that I can crowd worry out of mind by reading an absorbing book. When I was fiftynine, I had a prolonged nervous breakdown. During that period I began reading David Alec Wilson's monumental *Life of Carlyle*. It had a good deal to do with my convalescence because I became so absorbed in reading it that I forgot my despondency.

3 At another time when I was terribly depressed, I forced myself to become physically active almost every hour of the day. I played five or six sets of violent games of tennis every morning, then took a bath, had lunch, and played eighteen holes of golf every afternoon. On Friday night I danced until one o'clock in the morning. I am a great believer in working up a tremendous sweat. I found that depression and worry oozed out of my system with the sweat.

4 I learned long ago to avoid the folly of hurry, rush, and working under tension. I have always tried to apply the philosophy of Wilbur Cross. When he was Governor of Connecticut, he said to me: "Sometimes when I have too many things to do all at once, I sit down and relax and smoke my pipe for an hour and do nothing."

5 I have also learned that patience and time have a way of resolving our troubles. When I am worried about something, I try to see my troubles in their proper perspective. I say to myself: "Two months from now I shall not be worrying about this bad break, so why worry about it now? Why not assume now the same attitude that I will have two months from now?"

To sum up, here are the five ways in which Professor Phelps banished worry:

1 Live with gusto and enthusiasm: "I live every day as if it were the first day I had ever seen and the last I were going to see."

2 Read an interesting book: "When I had a prolonged nervous breakdown ... I began reading ... the Life of Carlyle ... and became so absorbed in reading it that I forgot my despondency."

3 Play games: "When I was terribly depressed, I forced myself to become physically active almost every hour of the day."

4 Relax while you work: "I long ago learned to avoid the folly of hurry, rush, and working under tension."

5 "I try to see my troubles in their proper perspective. I say to myself: 'Two months from now I shall not be worrying about this bad break, so why worry about it now? Why not assume now the same attitude that I will have two months from now?'"

✦ ✦ ✦

I Stood Yesterday. I Can Stand Today

By Dorothy Dix

I have been through the depths of poverty and sickness. When people ask me what has kept me going through the troubles that come to all of us, I always reply: "I stood yesterday. I can stand today. And I will not permit myself to think about what might happen tomorrow."

I have known want and struggle and anxiety and despair. I have always had to work beyond the limit of my strength. As I look back upon my life, I see it as a battlefield strewn with the wrecks of dead dreams and broken hopes and shattered illusions-a battle in which I always fought with the odds tremendously against me, and which has left me scarred and bruised and maimed and old before my time.

Yet I have no pity for myself; no tears to shed over the past and gone sorrows; no envy for the women who have been spared all I have gone through. For I have lived. They only existed. I have drank the cup of life down to its very dregs. They have only sipped the bubbles on top of it. I know things they will never know. I see things to which they are blind. It is only the women whose eyes have been washed clear with tears who get the broad vision that makes them little sisters to all the world.

I have learned in the great University of Hard Knocks a philosophy that no woman who has had an easy life ever acquires. I have learned to live each day as it comes and not to borrow trouble by dreading the morrow. It is the dark menace of the future that makes cowards of us. I put that dread from me because experience has taught me that when the time comes that I so fear, the strength and wisdom to meet it will be given me. Little annoyances no longer have the power to affect me. After you have seen your whole edifice of happiness topple and crash in ruins about you, it never matters to you again that a servant forgets to put the doilies under the finger bowls, or the cook spills the soup.

I have learned not to expect too much of people, and so I can still get happiness out of the friend who isn't quite true to me or the acquaintance who gossips. Above all, I have acquired a sense of humour, because there were so many things over which I had either to cry or laugh. And when a woman can joke over her troubles instead of having hysterics, nothing can ever hurt her much again. I do not regret the hardships I have known, because through them I have touched life at every point I have lived. And it was worth the price I had to pay.

Dorothy Dix conquered worry by living in "day-tight" compartments.

* * *

I Did Not Expect To Live To See The Dawn

By J.C. Penney

[On April 14, 1902, a young man with five hundred dollars in cash and a million dollars in determination opened a drygoods store in Kemmerer, Wyoming-a little mining town of a thousand people, situated on the old covered-wagon trail laid out by the Lewis and Clark Expedition. That young man and his wife lived in a half-storey attic above the store, using a large empty dry-goods box for a table and smaller boxes for chairs. The young wife wrapped her baby in a blanket and let it sleep under a counter while she stood beside it, helping her husband wait on customers. Today the largest chain of dry-goods stores in the world bears that man's name: the J.C. Penney stores-over sixteen hundred of them covering every state in the Union. I recently had dinner with Mr. Penney, and he told me about the most dramatic moment of his life.]

Years ago, I passed through a most trying experience. I was worried and desperate. My worries were not connected in any way whatever with the J. C. Penney Company. That business was solid and thriving; but I personally had made some unwise commitments prior to the crash of 1929. Like many other men, I was blamed for conditions for which I was in no way responsible. I was so harassed with worries that I couldn't sleep, and developed an extremely painful ailment known as shingles-a red rash and skin eruptions. I consulted a physician-a man with whom I had gone to high school as a boy in Hamilton, Missouri: Dr. Elmer Eggleston, a staff physician at the Kellogg Sanatorium in Battle Creek, Michigan. Dr. Eggleston put me to bed and warned me that I was a very ill man. A rigid treatment was prescribed. But nothing helped. I got weaker day by day. I was broken nervously and physically, filled with despair, unable to see even a ray of hope. I had nothing to live for. I felt I hadn't a friend left in the world, that even my family had turned against me. One night, Dr, Eggleston gave me a sedative, but the effect soon wore off and I awoke with an overwhelming conviction that this was my last night of life. Getting out of bed, I wrote farewell letters to my wife and to my son, saying that I did not expect to live to see the dawn.

When I awoke the next morning, I was surprised to find that I was still alive. Going downstairs, I heard singing in a little chapel where devotional exercises were held each morning. I can still remember the hymn they were singing: "God will take care of you." Going into the chapel, I listened with a weary heart to the singing, the reading of the Scripture lesson, and the prayer. Suddenly something happened. I can't explain it. I can only call it a miracle. I felt as if I had been instantly lifted out of the darkness of a dungeon into warm, brilliant sunlight. I felt as if I had been transported from hell to paradise. I felt the power of God as I had never felt it before. I realised then that I alone was responsible for all my troubles. I knew that God with His love was there to help me. From that day to this, my life has been free from worry. I am seventy-one years old, and the most dramatic and glorious twenty minutes of my life were those I spent in that chapel that morning: "God will take care of you."

J.C. Penney learned to overcome worry almost instantaneously, because he discovered the one perfect cure.

✢ ✢ ✢

I Go To The Gym To Punch The Bag Or Take A Hike Outdoors

By Colonel Eddie Eagan
New York Attorney, Rhodes Scholar Chairman, New York State Athletic Commission Former Olympic Light-Heavyweight Champion of the World

When I find myself worrying and mentally going round in endless circles like a camel turning a water wheel in Egypt, a good physical work-out helps me to chase those "blues" away. It may be running or a long hike in the country, or it may be a half-hour of bag punching or squash tennis at the gymnasium. Whichever it is, physical exercise clears my mental outlook. On a week-end I do a lot of physical sport, such as a run around the golf course, a game of paddle tennis, or a ski week-end in the Adirondacks. By my becoming physically tired, my mind gets a rest from legal problems, so that when I return to them, my mind has a new zest and power.

Quite often in New York, where I work, there is a chance for me to spend an hour at the Yale Club gym. No man can worry while he is playing squash tennis or skiing. He is too busy to worry. The large mental mountains of trouble become minute molehills that new thoughts and acts quickly smooth down.

I find the best antidote for worry is exercise. Use your muscles more and your brain less when you are worried, and you will be surprised at the result. It works that way with me-worry goes when exercise begins.

✦ ✦ ✦

I Was "The Worrying Wreck From Virginia Tech."

By Jim Birdsall
Plant Superintendent C.F. Muller Company 180 Baldwin Avenue, Jersey City, New Jersey

Seventeen years ago, when I was in military college at Blacksburg, Virginia, I was known as "the worrying wreck from Virginia Tech". I worried so violently that I often became ill. In fact, I was ill so often that I had a regular bed reserved for me at the college infirmary at all times. When the nurse saw me coming, she would run and give me a hypo. I worried about everything. Sometimes I even forgot what I was worrying about. I worried for fear I would be busted out of college because of my low grades. I had failed to pass my examinations in physics and other subjects, too. I knew I had to maintain an average grade of 75-84. I worried about my health, about my excruciating attacks of acute indigestion, about my insomnia. I worried about financial matters. I felt badly because I couldn't buy my girl candy or take her to dances as often as I wanted to. I worried for fear she would marry one of the other cadets. I was in a lather day and night over a dozen intangible problems.

In desperation, I poured out my troubles to Professor Duke Baird, professor of business administration at V.P.I.

The fifteen minutes that I spent with Professor Baird did more for my health and happiness than all the rest of the four years I spent in college. "Jim," he said, "you ought to sit down and face the facts. If you devoted half as much time and energy to solving your problems as you do to worrying about them, you wouldn't have any worries. Worrying is just a vicious habit you have learned."

He gave me three rules to break the worry habit:

Rule 1. Find out precisely what is the problem you are worrying about.

Rule 2. Find out the cause of the problem.

Rule 3. Do something constructive at once about solving the problem.

After that interview, I did a bit of constructive planning. Instead of worrying because I had failed to pass physics, I now asked myself why I had failed. I knew it wasn't because I was dumb, for I was editor-in-chief of The Virginia Tech Engineer.

I figured that I had failed physics because I had no interest in the subject. I had not applied myself because I couldn't see how it would help me in my work as an industrial engineer. But now I changed my attitude. I said to myself: "If the college authorities demand that I pass my physics examination before I obtain a degree, who am I to question their wisdom?"

So I enrolled for physics again. This time I passed because instead of wasting my time in resentment and worrying about how hard it was, I studied diligently.

I solved my financial worries by taking on some additional jobs, such as selling punch at the college dances, and by borrowing money from my father, which I paid back soon after graduation.

I solved my love worries by proposing to the girl that I feared might marry another cadet. She is now Mrs. Jim Birdsall.

As I look back at it now, I can see that my problem was one of confusion, a disinclination to find the causes of my worry and face them realistically.

Jim Birdsall learned to stop worrying because he ANALYSED his troubles. In fact, he used the very principles described in the chapter "How to Analyse and Solve Worry Problems."

✦ ✦ ✦

I Have Lived By This Sentence

By Dr. Joseph R. Sizoo
President, New Brunswick Theological Seminary (The oldest theological seminary in the United States, founded in 1784)

Years ago, in a day of uncertainty and disillusionment, when my whole life seemed to be overwhelmed by forces beyond my control, one morning quite casually I opened my New Testament and my eyes fell upon this sentence: "He that sent me is with me-the Father hath not left me alone." My life has never been the same since that hour. Everything for me has been forever different after that. I suppose that not a day has passed that I have not repeated it to myself. Many have come to me for counseling during these years, and I have always sent them away with this sustaining sentence. Ever since that hour when my eyes fell upon it, I have lived by this sentence. I have walked with it and I have found in it my peace and strength. To me it is the very essence of religion. It lies at the rock bottom of everything that makes life worth living. It is the Golden Text of my life.

✦ ✦ ✦

I Hit Bottom And Survived

By Ted Ericksen
16,237 South Cornuta Avenue, Bellflower, California
Southern California Representative National Enameling and Stamping Company

I used to be a terrible "worry wart". But no more. In the summer of 1942, I had an experience that banished worry from my life-for all time; I hope. That experience made every other trouble seem small by comparison.

For years I had wanted to spend a summer on a commercial fishing craft in Alaska, so in 1942 I signed on a thirty-two-foot salmon seining vessel out of Kodiak, Alaska. On a craft of this size, there is a crew of only three: the skipper who does the supervising, a No. 2 man who assists the skipper, and a general work horse, who is usually a Scandinavian. I am a Scandinavian.

Since salmon seining has to be done with the tides, I often worked twenty hours out of twenty-four. I kept up that schedule for a week at a time. I did everything that nobody else wanted to do. I washed the craft. I put away the gear. I cooked on a little wood-burning stove in a small cabin where the heat and fumes of the motor almost made me ill. I washed the dishes. I repaired the boat. I pitched the salmon from our boat into a tender that took the fish to a cannery. My feet were always wet in rubber boots. My boots were often filled with water, but I had no time to empty them. But all that was play compared to my main job, which was pulling what is called the "cork line". That operation simply means placing your feet on the stem of the craft and pulling in the corks and the webbing of the net. At least, that is what you are supposed to do. But, in reality, the net was so heavy that when I tried to pull it in, it wouldn't budge. What really happened was that in trying to pull in the cork line, I actually pulled in the boat. I pulled it along on my own power, since the net stayed where it was. I did all this for weeks on end. It was almost the end of me, too. I ached horribly. I ached all over. I ached for months.

When I finally did have a chance to rest, I slept on a damp lumpy mattress piled on top of the provisions locker. I would put one of the lumps in the mattress under the part of my back that hurt most-and sleep as if I had been dragged. I was drugged by complete exhaustion.

I am glad now that I had to endure all that aching and exhaustion because it has helped me stop worrying. Whenever I am confronted by a problem now-instead of worrying about it, I say to myself: "Ericksen, could this possibly be as bad as pulling the cork line?" And Ericksen invariably answers: "No, nothing could be that bad!" So I cheer up and tackle it with courage. I believe it is a good thing to have to endure an agonising experience occasionally. It is good to know that we have hit bottom and survived. That makes all our daily problems seem easy by comparison.

✦ ✦ ✦

I Used To Be One Of The World's Biggest Jackasses

By Percy H. Whiting
Managing Director, Dale Carnegie and Company 50 East 42nd Street, New York, New York

I have died more times from more different diseases than any other man, living, dead, or half dead.

I was no ordinary hypochondriac. My father owned a drug-store, and I was practically brought up in it. I talked to doctors and nurses every day, so I knew the names and symptoms of more and worse diseases than the average layman. I was no ordinary hypo-I had symptoms! I could worry for an hour or two over a disease and then have practically all the symptoms of a man who was suffering from it. I recall once that, in Great Barrington, Massachusetts, the town in which I lived, we had a rather severe diphtheria epidemic. In my father's drug-store, I had been selling medicines day after day to people who came from infected homes. Then the evil that I feared came upon me: I had diphtheria myself. I was positive I had it. I went to bed and worried myself into the standard symptoms. I sent for a doctor. He looked me over and said: "Yes, Percy, you've got it." That relieved my mind. I was never afraid of any disease when I had it. So, I turned over and went to sleep. The next morning I was in perfect health.

For years I distinguished myself and got a lot of attention and sympathy by specialising in unusual and fantastic disease-I died several times of both lockjaw and hydrophobia. Later on, I settled down to having the run-of-mill ailments-specialising on cancer and tuberculosis.

I can laugh about it now, but it was tragic then. I honestly and literally feared for years that I was walking on the edge of the grave. When it came time to buy a suit of clothes in the spring, I would ask myself: "Should I waste this money when I know I can't possibly live to wear this suit out?"

However, I am happy to report progress: in the past ten years, I haven't died even once.

How did I stop dying? By kidding myself out of my ridiculous imaginings. Every time I felt the dreadful symptoms coming on, I laughed at myself and said: "See here, Whiting, you have been dying from one fatal disease after another now for twenty years, yet you are in first-class health today. An insurance company recently accepted you for more insurance. Isn't it about time, Whiting, that you stood aside and had a good laugh at the worrying jackass you are?"

I soon found that I couldn't worry about myself and laugh at myself at one and the same time. So I've been laughing at myself ever since.

The point of this is: Don't take yourself too seriously. Try "just laughing" at some of your sillier worries, and see if you can't laugh them out of existence.

✦ ✦ ✦

I Have Always Tried To Keep My Line Of Supplies Open

By Gene Autry
The world's most famous and beloved singing cowboy

I figure that most worries are about family troubles and money. I was fortunate in marrying a smalltown Oklahoma girl who had the same background I had and enjoyed the same things. We both try to follow the golden rule, so we have kept our family troubles to a minimum.

I have kept my financial worries to a minimum also by doing two things. First, I have always followed a rule of absolute one hundred per cent integrity in everything. When I borrowed money, I paid back every penny. Few things cause more worry than dishonesty.

Second, when I started a new venture, I always kept on ace in the hole. Military experts say that the first principle of fighting a battle is to keep your line of supplies open. I figure that that principle applies to personal battles almost as much as to military battles. For example, as a lad down in Texas and Oklahoma, I saw some real poverty when the country was devastated by droughts. We had mighty hard scratching at times to make a living. We were so poor that my father used to drive across the country in a covered wagon with a string of horses and swap horses to make a living. I wanted something more reliable than that. So I got a job working for a railway-station agent and learned telegraphy in my spare time. Later, I got a job working as relief operator for the Frisco Railway. I was sent here, there, and yonder to relieve other station agents who were ill or on vacation or had more work than they could do. That job paid $150 per month. Later, when I started out to better myself, I always figured that that railroad job meant economic safety. So

I always kept the road open back to that job. It was my line of supplies, and I never cut myself off from it until I was firmly established in a new and better position.

For example, back in 1928, when I was working as a relief operator for the Frisco Railway in Chelsea, Oklahoma, a stranger drifted in one evening to send a telegram. He heard me playing the guitar and singing cowboy songs and told me I was good-told me that I ought to go to New York and get a job on the stage or radio. Naturally, I was flattered; and when I saw the name he signed to his telegram, I was almost breathless: Will Rogers.

Instead of rushing off to New York at once, I thought the matter over carefully for nine months. I finally came to the conclusion that I had nothing to lose and everything to gain by going to New York and giving the old town a whirl. I had a railroad pass: I could travel free. I could sleep sitting up in my seat, and I could carry some sandwiches and fruit for my meals.

So I went. When I reached New York, I slept in a furnished room for five dollars a week, ate at the Automat, and tramped the streets for ten weeks-and got nowhere. I would have been worried sick if I hadn't had a job to go back to. I had already worked for the railway five years. That meant I had seniority rights; but in order to protect those rights, I couldn't lay off longer than ninety days. By this time, I had already been in New York seventy days, so I rushed back to Oklahoma on my pass and began working again to protect my line of supply. I worked for a few months, saved money, and returned to New York for another try. This time I got a break. One day, while waiting for an interview in a recording-studio office, I played my guitar and sang a song to the girl receptionist: "Jeannine, I Dream of Lilac Time". While I was singing that song, the man who wrote it-Nat Schildkraut- drifted into the office. Naturally, he was pleased to hear anyone singing his song. So he gave me a note of introduction and sent me down to the Victor Recording Company. I made a record. I was no good-too stiff and self-conscious. So I took the advice of the Victor Recording man: I went back to Tulsa, worked for the railway by day, and at night I sang cowboy songs on a sustaining radio programme. I liked that arrangement. It meant that I was keeping my line of supplies open- so I had no worries.

I sang for nine months on radio station KVOO in Tulsa. During that time, Jimmy Long and I wrote a song entitled "That Silver-Haired Daddy of Mine". It caught on. Arthur Sattherly, head of the American Recording Company, asked me to make a recording. It clicked. I made a number of other recordings for fifty dollars each, and finally got a job singing cowboy songs over radio station WLS in Chicago. Salary: forty dollars a week. After singing there four years, my salary was raised to ninety dollars a week, and I picked up another three hundred dollars doing personal appearances every night in theatres.

Then in 1934, I got a break that opened up enormous possibilities. The League of Decency was formed to clean up the movies. So Hollywood producers decided to put on cowboy pictures; but they wanted a new kind of cowboy-one who could sing. The man who owned the American Recording Company was also part owner of Republic Pictures. "If you want a singing cowboy," he said to his associates, "I have got one making records for us." That is how I broke into the movies. I started making singing-cowboy pictures for one hundred dollars a week. I had serious doubts about whether I would succeed in pictures, but I didn't worry. I knew I could always go back to my old job.

My success in pictures exceeded my wildest expectations. I now get a salary of one hundred thousand a year plus one half of all the profits on my pictures. However, I realise that this arrangement won't go on forever. But I am not worried. I know that no matter what happens-even if I lose every dollar I have-I can always go back to Oklahoma and get a job working for the Frisco Railway. I have protected my line of supplies.

✦ ✦ ✦

I Heard A Voice In India

By E. Stanley Jones
One of America's most dynamic speakers and the most famous missionary of his generation

I have devoted forty years of my life to missionary work in India. At first, I found it difficult to endure the terrible heat plus the nervous strain of the great task that stretched before me. At the end of eight years, I was suffering so severely from brain fatigue and nervous exhaustion that I collapsed, not once but several times. I was ordered to take a year's furlough in America. On the boat returning to America, I collapsed again while speaking at a Sunday-morning service on the ship, and the ship's doctor put me to bed for the remainder of the trip.

After a year's rest in America, I started back to India, but stopped on the way to hold evangelistic meetings among the university students in Manila. In the midst of the strain of these meetings, I collapsed several times. Physicians warned me that if I returned to India, I would die. In spite of their warnings, I continued on to India, but I went with a deepening cloud upon me. When I arrived in Bombay, I was so broken that I went straight to the hills and rested for several months. Then I returned to the plains to continue my work. It was no use. I collapsed and was forced to return to the hills for another long rest. Again I descended to the plains, and again I was shocked and crushed to discover that I couldn't take it. I was exhausted mentally, nervously, and physically. I was completely at the end of my resources. I feared that I would be a physical wreck for the balance of my life.

If I didn't get help from somewhere, I realised that I would have to give up my missionary career, go back to America, and work on a farm to try to regain my health. It was one of my darkest hours. At that time I was holding a series of meetings in Lucknow. While praying one night, an event happened that completely transformed my life. While in prayer-and I was not particularly thinking about myself at the time-a voice seemed to say: "Are you yourself ready for this work to which I have called you?"

I replied: "No, Lord, I am done for. I have reached the end of my resources."

The Voice replied "If you will turn that over to Me and not worry about it, I will take care of it."

I quickly answered: "Lord, I close the bargain right here."

A great peace settled into my heart and pervaded my whole being. I knew it was done! Life-abundant life-had taken possession of me. I was so lifted up that I scarcely touched the road as I quietly walked home that night. Every inch was holy ground. For days after that I hardly knew I had a body. I went through the days, working all day and far into the night, and came down to bedtime wondering why in the world I should ever go to bed at all, for there was not the slightest trace of tiredness of any kind. I seemed possessed by life and peace and rest-by Christ Himself.

The question came as to whether I should tell this. I shrank from it, but I felt I should-and did. After that it was sink or swim before everybody. More than a score of the most strenuous years of my life have gone by since then, but the old trouble has never returned. I have never had such health. But it was more than a physical touch. I seemed to have tapped new life for body, mind, and spirit. After that experience, life for me functioned on a permanently higher level. And I had done nothing but take it!

During the many years that have gone by since then, I have travelled all over the world, frequently lecturing three times a day, and have found time and strength to write *The Christ of the Indian Road* and eleven other books. Yet in the midst of all this, I have never missed, or even been late to, an appointment. The worries that once beset me have long since vanished, and now, in my sixty-third year, I am overflowing with abounding vitality and the joy of serving and living for others.

I suppose that the physical and mental transformation that I have experienced could be picked to pieces psychologically and explained. It does not matter. Life is bigger than processes and overflows and dwarfs them.

This one thing I know: my life was completely transformed and uplifted that night in Lucknow, thirtyone years ago, when at the depth of my weakness and depression, a voice said to me: "If you will turn that over to Me and not worry about it, I will take care of it," and I replied: "Lord, I close the bargain right here."

✦ ✦ ✦

When The Sheriff Came In My Front Door

By Homer Croy
Novelist, 150 Pinehurst Avenue, New York, New York

The bitterest moment of my life occurred one day in 1933 when the sheriff came in the front door and I went out the back. I had lost my home at 10 Standish Road, Forest Hills, Long Island, where my children were born and where I and my family had lived for eighteen years. I had never dreamed that this could happen to me. Twelve years before, I thought I was sitting on top of the world. I had sold the motion-picture rights to my novel *West of the Water Tower* for a top Hollywood price. I lived abroad with my family for two years. We summered in Switzerland and wintered on the French Riviera- just like the idle rich.

I spent six months in Paris and wrote a novel entitled *They Had to See Paris*. Will Rogers appeared in the screen version. It was his first talking picture. I had tempting offers to remain in Hollywood and write several of Will Rogers' pictures. But I didn't. I returned to New York. And my troubles began!

It slowly dawned on me that I had great dormant abilities that I had never developed. I began to fancy myself a shrewd businessman. Somebody told me that John Jacob Astor had made millions investing in vacant land in New York. Who was Astor? Just an immigrant peddler with an accent. If he could do it, why couldn't I? ... I was going to be rich! I began to read the yachting magazines.

I had the courage of ignorance. I didn't know any more about buying and selling real estate than an Eskimo knows about oil furnaces. How was I to get the money to launch myself on my spectacular financial career? That was simple. I mortgaged my home, and bought some of the finest building lots in Forest Hills. I was going to hold this land until it reached a fabulous price, then sell it and live in luxury-I who had never sold a piece of real estate as big as a doll's handkerchief. I pitied the plodders who slaved in offices for a mere salary. I told myself that God had not seen fit to touch every man with the divine fire of financial genius.

Suddenly, the great depression swept down upon me like a Kansas cyclone and shook me as a tornado would shake a hen coop.

I had to pour $220 a month into that monster-mouthed piece of Good Earth. Oh, how fast those months came! In addition, I had to keep up the payments on our now-mortgaged house and find enough food. I was worried. I tried to write humour for the magazines. My attempts at humour sounded like the lamentations of Jeremiah! I was unable to sell anything. The novel I wrote failed. I ran out of money. I had nothing on which I could borrow money except my typewriter and the gold fillings in my teeth. The milk company stopped delivering milk. The gas company turned off the gas. We had to buy one of those little outdoor camp stoves you see advertised; it had a cylinder of gasoline; you pump it up by hand and it shoots out a flame with a hissing like an angry goose.

We ran out of coal; the company sued us. Our only heat was the fireplace. I would go out at night and pick up boards and left-overs from the new homes that the rich people were building ... I who had started out to be one of these rich people.

I was so worried I couldn't sleep. I often got up in the middle of the night and walked for hours to exhaust myself so I could fall asleep.

I lost not only the vacant land I had bought, but all my heart's blood that I had poured into it.

The bank closed the mortgage on my home and put me and my family out on the street.

In some way, we managed to get hold of a few dollars and rent a small apartment. We moved in the last day of 1933. I sat down on a packing case and looked around. An old saying of my mother's came back: "Don't cry over spilt milk."

But this wasn't milk. This was my heart's blood!

After I had sat there a while I said to myself: "Well, I've hit bottom and I've stood it. There's no place to go now but up."

I began to think of the fine things that the mortgage had not taken from me. I still had my health and my friends. I would start again. I would not grieve about the past. I would repeat to myself every day the words I had often heard my mother say about spilt milk.

I put into my work the energy that I had been putting into worrying. Little by little, my situation began to improve. I am almost thankful now that I had to go through all that misery; it gave me strength, fortitude, and confidence. I know now what it means to hit bottom. I know it doesn't kill you. I know we can stand more than we think we can. When little worries and anxieties and uncertainties try to disturb me now, I banish them by reminding myself of the time I sat on the packing case and said: "I've hit bottom and I've stood it. There is no place to go now but up."

What's the principle here? Don't try to saw sawdust. Accept the inevitable! If you can't go lower, you can try going up.

✣ ✣ ✣

The Toughest Opponent I Ever Fought Was Worry

By Jack Dempsey

During my career in the ring, I found that Old Man Worry was an almost tougher opponent than the heavyweight boxers I fought. I realised that I had to learn to stop worrying, or worry would sap my vitality and undermine my succcss. So, little by little, I worked out a system for myself. Here are some of the things I did:

1. To keep up my courage in the ring, I would give myself a pep talk during the fight. For example, while I was fighting Firpo, I kept saying over and over: "Nothing is going to stop me. He is not going to hurt me. I won't feel his blows. I can't get hurt. I am going to keep going, no matter what happens." Making positive statements like that to myself, and thinking positive thoughts, helped me a lot. It even kept my mind so occupied that I didn't feel the blows. During my career, I have had my lips smashed, my eyes cut, my ribs cracked-and Firpo knocked me clear through the ropes, and I landed on a reporter's typewriter and wrecked it. But I never felt even one of Firpo's blows. There was only one blow that I ever really felt. That was the night Lester Johnson broke three of my ribs. The punch never hurt me; but it affected my breathing. I can honestly say I never felt any other blow I ever got in the ring.

2. Another thing I did was to keep reminding myself of the futility of worry. Most of my worrying was done before the big bouts, while I was going through training. I would often lie awake at nights for hours, tossing and worrying, unable to sleep. I would worry for fear I might break my hand or sprain my ankle or get my eye cut badly in the first round so I couldn't

co-ordinate my punches. When I got myself into this state of nerves, I used to get out of bed, look into the mirror, and give myself a good talking to. I would say: "What a fool you are to be worrying about something than hasn't happened and may never happen. Life is short. I have only a few years to live, so I must enjoy life." I kept saying to myself: "Nothing is important but my health. Nothing is important but my health." I kept reminding myself that losing sleep and worrying would destroy my health. I found that by saying these things to myself over and over, night after night, year after year, they finally got under my skin, and I could brush off my worries like so much water.

3. The third-and best-thing I did was pray! While I was training for a bout, I always prayed several times a day. When I was in the ring, I always prayed just before the bell sounded for each round. That helped me fight with courage and confidence. I have never gone to bed in my life without saying a prayer; and I have never eaten a meal in my life without first thanking God for it ... Have my prayers been answered? Thousands of times!

✦ ✦ ✦

I Prayed To God To Keep Me Out Of An Orphan's Home

By Kathleen Halter
Housewife, 1074 Roth, University City 14, Missouri

As a little child, my life was filled with horror. My mother had heart trouble. Day after day, I saw her faint and fall to the floor. We all feared she was going to die, and I believed that all little girls whose mothers died were sent to the Central Wesleyan Orphans' Home, located in the little town of Warrenton, Missouri, where we lived. I dreaded the thought of going there, and when I was six years old I prayed constantly: "Dear God, please let my mummy live until I am old enough not to go to the orphans' home."

Twenty years later, my brother, Meiner, had a terrible injury and suffered intense pain until he died two years later. He couldn't feed himself or turn over in bed. To deaden his pain, I had to give him morphine hypodermics every three hours, day and night. I did this for two years. I was teaching music at the time at the Central Wesleyan College in Warrenton, Missouri. When the neighbours heard my brother screaming with pain, they would telephone me at college and I would leave my music class and rush home to give my brother another injection of morphine. Every night when I went to bed, I would set the alarm clock to go off three hours later so I would be sure to get up to attend to my brother. I remember that on winter nights I would keep a bottle of milk outside the window, where it would freeze and turn into a kind of ice-cream that I loved to eat. When the alarm went off, this ice cream outside the window gave me an additional incentive to get up.

In the midst of all these troubles, I did two things that kept me from indulging in self-pity and worrying and embittering my life with resentment. First, I kept myself busy teaching music from twelve to fourteen hours a day, so I had little time to think of my troubles; and when I was tempted to feel sorry for myself, I kept saying to myself over and over: "Now, listen, as long as you can walk and feed yourself and are free from intense pain, you ought to be the happiest person in the world. No matter what happens, never forget that as long as you live! Never! Never!"

I was determined to do everything in my power to cultivate an unconscious and continuous attitude of gratefulness for my many blessings. Every morning when I awoke, I would thank God that conditions were no worse than they were; and I resolved that in spite of my troubles I would be the happiest person in Warrenton, Missouri. Maybe I didn't succeed in achieving that goal, but I did succeed in making myself the most grateful young woman in my town-and probably few of my associates worried less than I did.

This Missouri music teacher applied two principles described in this book: she kept too busy to worry, and she counted her blessings. The same technique may be helpful to you.

✦ ✦ ✦

I Was Acting Like An Hysterical Woman

By Cameron Shipp
Magazine Writer

I had been working very happily in the publicity department of the Warner Brothers studio in California for several years. I was a unit man and feature writer. I wrote stories for newspapers and magazines about Warner Brother stars.

Suddenly, I was promoted. I was made the assistant publicity director. As a matter of fact, there was a change of administrative policy, and I was given an impressive title: Administrative Assistant.

This gave me an enormous office with a private refrigerator, two secretaries, and complete charge of a staff of seventy-five writers, exploiters, and radio men. I was enormously impressed. I went straight out and bought a new suit. I tried to speak with dignity. I set up filing systems, made decisions with authority, and ate quick lunches.

I was convinced that the whole public-relations policy of Warner Brothers had descended upon my shoulders. I perceived that the lives, both private and public, of such renowned persons as Bette Davis, Olivia De Havilland, James Cagney, Edward G. Robinson, Errol Flynn, Humphrey Bogart, Ann Sheridan, Alexis Smith, and Alan Hale were entirely in my hands.

In less than a month I became aware that I had stomach ulcers. Probably cancer.

My chief war activity at that time was chairman of the War Activities Committee of the Screen Publicists Guild. I liked to do this work, liked to meet my friends at guild meetings. But these gatherings became matters of dread. After every meeting, I was violently ill. Often I had to stop my car on the way home, pulling myself together before I could drive on. There seemed to be so much to do, so little time in which to do it. It was all vital. And I was woefully inadequate.

I am being perfectly truthful-this was the most painful illness of my entire life. There was always a tight fist in my vitals. I lost weight. I could not sleep. The pain was constant.

So I went to see a renowned expert in internal medicine. An advertising man recommended him. He said this physician had many clients who were advertising men.

This physician spoke only briefly, just enough for me to tell him where I hurt and what I did for a living. He seemed more interested in my job than in my ailments, but I was soon reassured: for two weeks, daily, he gave me every known test. I was probed, explored, X-rayed, and fluoroscoped. Finally, I was instructed to call on him and hear the verdict.

"Mr. Shipp," he said, leaning back and offering me a cigarette, "we have been through these exhaustive tests. They were absolutely necessary, although I knew of course after my first quick examination that you did not have stomach ulcers.

"But I knew, because you are the kind of man you are and because you do the kind of work you do, that you would not believe me unless I showed you. Let me show you."

So he showed me the charts and the X-rays and explained them. He showed me I had no ulcers.

"Now," said the doctor, "this costs you a good deal of money, but it is worth it to you. Here is the prescription: don't worry.

"Now"-he stopped me as I started to expostulate-"now, I realise that you can't follow the prescription immediately, so I'll give you a crutch. Here are some pills. They contain belladonna. Take as many as you like. When you use these up, come back and I'll give you more. They won't hurt you. But they will always relax you.

"But remember: you don't need them. All you have to do is quit worrying.

"If you do start worrying again, you'll have to come back here and I'll charge you a heavy fee again. How about it?"

I wish I could report that the lesson took effect that day and that I quit worrying immediately. I didn't. I took the pills for several weeks, whenever I felt a worry coming on. They worked. I felt better at once.

But I felt silly taking these pills. I am a big man physically. I am almost as tall as Abe Lincoln was and I weigh almost two hundred pounds. Yet here I was taking little white pills to relax myself. I was acting like an hysterical woman. When my friends asked me why I was taking pills, I was ashamed to tell the truth. Gradually I began to laugh at myself. I said: "See here, Cameron Shipp, you are acting like a fool. You are taking yourself and your little activities much, much too seriously. Bette Da vis and James Cagney and Edward G. Robinson were world-famous before you started to handle their publicity; and if you dropped dead tonight, Warner Brothers and their stars would manage to get along without you. Look at Eisenhower, General Marshall, MacArthur, Jimmy Doolittle and Admiral Kingthey are running the war without taking pills. And yet you can't serve as chairman of the War Activities Committee of the Screen Publicists Guild without taking little white pills to keep your stomach from twisting and turning like a Kansas whirlwind."

I began to take pride in getting along without the pills. A little while later, I threw the pills down the drain and got home each night in time to take a little nap before dinner and gradually began to lead a normal life. I have never been back to see that physician.

But I owe him much, much more than what seemed like a stiff fee at the time. He taught me to laugh at myself. But I think the really skilful thing he did was to refrain from laughing at me, and to refrain from telling me I had nothing to worry about. He took me seriously. He saved my face. He gave me an out in a small box. But he knew then, as well as I know now, that the cure wasn't in those silly little pills-the cure was in a change in my mental attitude.

The moral of this story is that many a man who is now taking pills would do better to read Chapter 7, and relax.

✣ ✣ ✣

I Learned To Stop Worrying By Watching My Wife Wash Dishes

By Reverend William Wood
204 Hurlbert Street, Charlevoix, Michigan

A few years ago, I was suffering intensely from pains in my stomach. I would awaken two or three times each night, unable to sleep because of these terrific pains. I had watched my father die from cancer of the stomach, and I feared that I too had a stomach cancer-or, at least, stomach ulcers. So I went to Byrne's Clinic at Petosky, Michigan, for an examination. Dr. Lilga, a stomach specialist, examined me with a fluoroscope and took an X-ray of my stomach. He gave me medicine to make me sleep and assured me that I had no stomach ulcers or cancer. My stomach pains, he said, were caused by emotional strains. Since I am a minister, one of his first questions was: "Do you have an old crank on your church board?"

He told me what I already knew; I was trying to do too much. In addition to my preaching every Sunday and carrying the burdens of the various activities of the church, I was also chairman of the Red Cross, president of the Kiwanis. I also conducted two or three funerals each week and a number of other activities.

I was working under constant pressure. I could never relax. I was always tense, hurried, and highstrung. I got to the point where I worried about everything. I was living in a constant dither. I was in such pain that I gladly acted on Dr. Lilga's advice. I took Monday off each week, and began eliminating various responsibilities and activities.

One day while cleaning out my desk, I got an idea that proved to be immensely helpful. I was looking over an accumulation of old notes on sermons and other memos on matters that were now past and gone. I crumpled them up one by one and tossed them into the wastebasket. Suddenly I stopped and said to myself: "Bill, why don't you do the same thing with your worries that you are doing with these notes? Why don't you crumple up your worries about yesterday's problems and toss them into the wastebasket?" That one idea gave me immediate inspiration-gave me the feeling of a weight being lifted from my shoulders. From that day to this, I have made it a rule to throw into the wastebasket all the problems that I can no longer do anything about.

Then, one day while wiping the dishes as my wife washed them, I got another idea. My wife was singing as she washed the dishes, and I said to myself: "Look, Bill, how happy your wife is. We have been married eighteen years, and she has been washing dishes all that time. Suppose when we got married she had looked ahead and seen all the dishes she would have to wash during those eighteen years that stretched ahead. That pile of dirty dishes would be bigger than a barn. The very thought of it would have appalled any woman."

Then I said to myself: "The reason my wife doesn't mind washing the dishes is because she washes only one day's dishes at a time." I saw what my trouble was. I was trying to wash today's dishes, yesterday's dishes and dishes that weren't even dirty yet.

I saw how foolishly I was acting. I was standing in the pulpit, Sunday mornings, telling other people how to live, yet, I myself was leading a tense, worried, hurried existence. I felt ashamed of myself.

Worries don't bother me any more now. No more stomach pains. No more insomnia. I now crumple up yesterday's anxieties and toss them into the wastebasket, and I have ceased trying to wash tomorrow's dirty dishes today.

Do you remember a statement quoted earlier in this book? "The load of tomorrow, added to that of yesterday, carried today, makes the strongest falter." ... Why even try it?

+ + +

I Found The Answer-keep Busy!

By Del Hughes
Public Accountant, 607 South Euclid Avenue, Bay City, Michigan

In 1943 I landed in a veterans' hospital in Albuquerque, New Mexico, with three broken ribs and a punctured lung. This had happened during a practice Marine amphibious landing off the Hawaiian Islands. I was getting ready to jump off the barge, on to the beach, when a big breaker swept in, lifted the barge, and threw me off balance and smashed me on the sands. I fell with such force that one of my broken ribs punctured my right lung.

After spending three months in the hospital, I got the biggest shock of my life. The doctors told me that I showed absolutely no improvement. After some serious thinking, I figured that worry was preventing me from getting well. I had been used to a very active life, and during these three months I had been flat on my back twenty-four hours a day with nothing to do but think. The more I thought, the more I worried: worried about whether I would ever be able to take my place in the world. I worried about whether I would remain a cripple the rest of my life, and about whether I would ever be able to get married and live a normal life.

I urged my doctor to move me up to the next ward, which was called the "Country Club" because the patients were allowed to do almost anything they cared to do.

In this "Country Club" ward, I became interested in contract bridge. I spent six weeks learning the game, playing bridge with the other fellows, and reading Culbertson's books on bridge. After six weeks, I was playing nearly every evening for the rest of my stay in the hospital. I also became interested in painting with oils, and I studied this art under an instructor every afternoon from three to five. Some of my paintings were so good that you could almost tell what they were! I also tried my hand at soap and wood carving, and read a number of books on the subject and found it fascinating. I kept myself so busy that I had no time to worry about my physical condition. I even found time to read books on psychology given to me by the Red Cross. At the end of three months, the entire medical staff came to me and congratulated me on "making an amazing improvement". Those were the sweetest words I had ever heard since the days I was born. I wanted to shout with joy.

The point I am trying to make is this: when I had nothing to do but lie on the flat of my back and worry about my future, I made no improvement whatever. I was poisoning my body with worry. Even the broken ribs couldn't heal. But as soon as I got my mind off myself by playing contract bridge, painting oil pictures, and carving wood, the doctors declared I made "an amazing improvement".

I am now leading a normal healthy life, and my lungs are as good as yours.

Remember what George Bernard Shaw said? "The secret of being miserable is to have the leisure to bother about whether you are happy or not." Keep active, keep busy!

✦ ✦ ✦

Time Solves A Lot Of Things

By Louis T. Montant, Jr.
Sales and Market Analyst, 114 West 64th Street, New York, New York

Worry caused me to lose ten years of my life. Those ten years should have been the most fruitful and richest years of any young man's life-the years from eighteen to twenty-eight.

I realise now that losing those years was no one's fault but my own.

I worried about everything: my job, my health, my family, and my feeling of inferiority. I was so frightened that I used to cross the street to avoid meeting people I knew. When I met a friend on the street, I would often pretend not to notice him, because I was afraid of being snubbed.

I was so afraid of meeting strangers-so terrified in their presence-that in one space of two weeks I lost out on three different jobs simply because I didn't have the courage to tell those three different prospective employers what I knew I could do.

Then one day eight years ago, I conquered worry in one afternoon-and have rarely worried since then. That afternoon I was in the office of a man who had had far more troubles than I had ever faced, yet he was one of the most cheerful men I had ever known. He had made a fortune in 1929, and lost every cent. He had made another fortune in 1933, and lost that; and another fortune in 1937, and lost that, too. He had gone through bankruptcy and had been hounded by enemies and creditors. Troubles that would have broken some men and driven them to suicide rolled off him like water off a duck's back.

As I sat in his office that day eight years ago, I envied him and wished that God had made me like him.

As we were talking, he tossed a letter to me that he had received that morning and said: "Read that."

It was an angry letter, raising several embarrassing questions. If I had received such a letter, it would have sent me into a tail-spin. I said: "Bill, how are you going to answer it?"

"Well," Bill said, "I'll tell you a little secret. Next time you've really got something to worry about, take a pencil and a piece of paper, and sit down and write out in detail just what's worrying you. Then put that piece of paper in the lower right-hand drawer of your desk. Wait a couple of weeks, and then look at it. If what you wrote down still worries you when you read it, put that piece of paper back in your lower right-hand drawer. Let it sit there for another two weeks. It will be safe there. Nothing will happen to it. But in the meantime, a lot may happen to the problem that is worrying you. I have found that, if only I have patience, the worry that is trying to harass me will often collapse like a pricked balloon."

That bit of advice made a great impression on me. I have been using Bill's advice for years now, and, as a result, I rarely worry about anything.

Times solves a lot of things. Time may also solve what you are worrying about today.

✦ ✦ ✦

I Was Warned Not To Try To Speak Or To Move Even A Finger

By Joseph L. Ryan
Supervisor, Foreign Division, Royal Typewriter Company 51 Judson Place, Rockville Centre, Long Island, New York

Several years ago I was a witness in a lawsuit that caused me a great deal of mental strain and worry. After the case was over, and I was returning home in the train, I had a sudden and violent physical collapse. Heart trouble. I found it almost impossible to breathe.

When I got home the doctor gave me an injection. I wasn't in bed-I hadn't been able to get any farther than the living-room settee. When I regained consciousness, I saw that the parish priest was already there to give me final absolution!

I saw the stunned grief on the faces of my family. I knew my number was up. Later, I found out that the doctor had prepared my wife for the fact that I would probably be dead in less than thirty minutes. My heart was so weak I was warned not to try to speak or to move even a finger.

I had never been a saint, but I had learned one thing-not to argue with God. So I closed my eyes and said: "Thy will be done. ... If it has to come now, Thy will be done."

As soon as I gave in to that thought, I seemed to relax all over. My terror disappeared, and I asked myself quickly what was the worst that could happen now. Well, the worst seemed to be a possible return of the spasms, with excruciating pains- then all would be over. I would go to meet my Maker and soon be at peace.

I lay on that settee and waited for an hour, but the pains didn't return. Finally, I began to ask myself what I would do with my life if I didn't die now. I determined that I would exert every effort to regain my health. I would stop abusing myself with tension and worry and rebuild my strength.

That was four years ago. I have rebuilt my strength to such a degree that even my doctor is amazed at the improvement my cardiograms show. I no longer worry. I have a new zest for life. But I can honestly say that if I hadn't faced the worst- my imminent death-and then tried to improve upon it, I don't believe I would be here today. If I hadn't accepted the worst, I believe I would have died from my own fear and panic.

Mr. Ryan is alive today because he made use of the principle described in the Magic Formula-FACE THE WORST THAT CAN HAPPEN.

✦ ✦ ✦

I Am A Great Dismisser

By Ordway Tead
Chairman of the Board of Higher Education New York, New York

WORRY is a habit-a habit that I broke long ago. I believe that my habit of refraining from worrying is due largely to three things.

First: I am too busy to indulge in self-destroying anxiety. I have three main activities-each one of which should be virtually a full-time job in itself. I lecture to large groups at Columbia University: I am also chairman of the Board of Higher Education of New York City. I also have charge of the Economic and Social Book Department of the publishing firm of Harper and Brothers. The insistent demands of these three tasks leave me no time to fret and stew and run around in circles.

Second: I am a great dismisser. When I turn from one task to another, I dismiss all thoughts of the problems I had been thinking about previously. I find it stimulating and refreshing to turn from one activity to another. It rests me. It clears my mind.

Third: I have had to school myself to dismiss all these problems from my mind when I close my office desk. They are always continuing. Each one always has a set of unsolved problems demanding my attention. If I carried these issues home with me each night, and worried about them, I would destroy my health; and, in addition, I would destroy all ability to cope with them.

Ordway Tead is a master of the Four Good Working Habits. Do you remember what they are?

If I Had Not Stopped Worrying, I Would Have Been In My Grave Long Ago

By Connie Mack

I have been in professional baseball for over sixty-three years. When I first started, back in the eighties, I got no salary at all. We played on vacant lots, and stumbled over tin cans and discarded horse collars. When the game was over, we passed the hat. The pickings were pretty slim for me, especially since I was the main support of my widowed mother and my younger brothers and sisters. Sometimes the ball team would have to put on a strawberry supper or a clambake to keep going. I have had plenty of reason to worry. I am the only baseball manager who ever finished in last place for seven consecutive years. I am the only manager who ever lost eight hundred games in eight years. After a series of defeats, I used to worry until I could hardly eat or sleep. But I stopped worrying twenty-five years ago, and I honestly believe that if I hadn't stopped worrying then, I would have been in my grave long ago.

As I looked back over my long life (I was born when Lincoln was President), I believe I was able to conquer worry by doing these things:

1. I saw how futile it was. I saw it was getting me nowhere and was threatening to wreck my career.

2. I saw it was going to ruin my health.

3. I kept myself so busy planning and working to win games in the future that I had no time to worry over games that were already lost.

4. I finally made it a rule never to call a player's attention to his mistakes until twenty-four hours after the game. In my early

days, I used to dress and undress with the players. If the team had lost, I found it impossible to refrain from criticising the players and from arguing with them bitterly over their defeats. I found this only increased my worries. Criticising a player in front of the others didn't make him want to co-operate. It really made him bitter. So, since I couldn't be sure of controlling myself and my tongue immediately after a defeat, I made it a rule never to see the players right after a defeat. I wouldn't discuss the defeat with them until the next day. By that time, I had cooled off, the mistakes didn't loom so large, and I could talk things over calmly and the men wouldn't get angry and try to defend themselves.

5. I tried to inspire players by building them up with praise instead of tearing them down with fault finding. I tried to have a good word for everybody.

6. I found that I worried more when I was tired; so I spend ten hours in bed every night, and I take a nap every afternoon. Even a five-minute nap helps a lot.

7. I believe I have avoided worries and lengthened my life by continuing to be active. I am eighty-five, but I am not going to retire until I begin telling the same stories over and over. When I start doing that, I'll know then that I am growing old.

Connie Mack never read a book on HOW TO STOP WORRYING so he made out his own roles. Why don't YOU make a list of the rules you have found helpful in the past-and write them out here?

Ways I Have Found Helpful in Overcoming Worry:

1 ____________________
2 ____________________
3 ____________________
4 ____________________

✣ ✣ ✣

One At A Time Gentleman, One At A Time

By John Homer Miller
Author of Take a Look at Yourself

I Discovered years ago that I could not escape my worries by trying to ran away from them, but that I could banish them by changing my mental attitude toward them. I discovered that my worries were not outside but inside myself.

As the years have gone by, I have found that time automatically takes care of most of my worries. In fact, I frequently find it difficult to remember what I was worrying about a week ago. So I have a rule: never to fret over a problem until it is at least a week old. Of course, I can't always put a problem completely out of mind for a week at a time, but I can refuse to allow it to dominate my mind until the allotted seven days have passed, either the problem has solved itself or I have so changed my mental attitude that it no longer has the power to trouble me greatly.

I have been greatly helped by reading the philosophy of Sir William Osier, a man who was not only a great physician, but a great artist in the greatest of all arts: the art of living. One of his statements has helped me immensely in banishing worries. Sir William said, at a dinner given in his honour: "More than to anything else, I owe whatever success I have had to the power of settling down to the day's work and trying to do it well to the best of my ability and letting the future take care of itself."

In handling troubles, I have taken as my motto the words of an old parrot that my father used to tell me about. Father told me of a parrot that was kept in a cage hanging over the doorway in a hunting club in Pennsylvania. As the members of the club passed through the door, the parrot repeated over and over the only words he knew: "One at a time, gentlemen, one at a time."

Father taught me to handle my troubles that way: "One at a time, gentlemen, one at a time." I have found that taking my troubles one at a time has helped me to maintain calm and composure amidst pressing duties and unending engagements. "One at a time, gentlemen, one at a time."

Here again, we have one of the basic principles in conquering worry: LIVE IN DAY-TIGHT COMPARTMENTS. Why don't you turn back and read that chapter again?

✦✦✦

I Now Look For The Green Light

By Joseph M. Cotter
1534 Fargo Avenue, Chicago, Illinois

From the time I was a small boy, throughout the early stages of young manhood, and during my adult life, I was a professional worrier. My worries were many and varied. Some were real; most of them were imaginary. Upon rare occasions I would find myself without anything to worry about-then I would worry for fear I might be overlooking something.

Then, two years ago, I started out on a new way of living. This required making a self-analysis of my faults-and a very few virtues-a "searching and fearless moral inventory" of myself. This brought out clearly what was causing all this worry.

The fact was that I could not live for today alone. I was fretful of yesterday's mistakes and fearful of the future.

I was told over and over that "today was the tomorrow I had worried about yesterday". But it wouldn't work on me. I was advised to live on a twenty-four-hour programme. I was told that today was the only day over which I had any control and that I should make the most of my opportunities each day. I was told that if I did that, I would be so busy I would have no time to worry about any other day-past or future. Advise was logical, but somehow I found it hard to put these darned ideas to work for me.

Then like a shot from out of the dark, I found the answer-and where do you suppose I found it? On a North-western Railroad platform at seven p.m. on May 31, 1945. It was an important hour for me. That is why I remember it so clearly.

We were taking some friends to the train. They were leaving on The City of Los Angeles, a streamliner, to return from a vacation. War was still on-crowds were heavy that year. Instead of boarding the train with my wife, I wandered down the tracks towards the front of the train. I stood looking at the big shiny engine for a minute. Presently I looked down the track and saw a huge semaphore. An amber light was showing. Immediately this light turned to a bright green. At that moment, the engineer started clanging a bell; I heard the familiar "All aboard!" and, in a matter of seconds, that huge streamliner began to move out of that station on its 2,300-mile trip.

My mind started spinning. Something was trying to make sense to me. I was experiencing a miracle. Suddenly it dawned on me. The engineer had given me the answer I had been seeking. He was starting out on that long journey with only one green light to go by. If I had been in his place, I would want to see all the green lights for the entire journey. Impossible, of course, yet that was exactly what I was trying to do with my life-sitting in the station, going no place, because I was trying too hard to see what was ahead for me.

My thoughts kept coming. That engineer didn't worry about trouble that he might encounter miles ahead. There probably would be some delays, some slowdowns, but wasn't that why they had signal systems ? Amber lights-reduce speed and take it easy. Red lights-real danger up ahead-stop. That was what made train travel safe. A good signal system.

I asked myself why I didn't have a good signal system for my life. My answer was-I did have one. God had given it to me. He controls it, so it has to be foolproof. I started looking for a green light. Where could I find it? Well, if God created the green lights, why not ask Him? I did just that.

And now by praying each morning, I get my green light for that day. I also occasionally get amber lights that slow me down. Sometimes I get red lights that stop me before I crack up. No more worrying for me since that day two years ago when I made this discovery. During those two years, over seven hundred green lights have shown for me, and the trip through life is so much easier without the worry of what colour the next light will be. No matter what colour it may be, I will know what to do.

✦ ✦ ✦

How John D. Rockefeller Lived on Borrowed Time for Forty-five Years

John D. Rockefeller, Sr., had accumulated his first million at the age of thirty-three. At the age of fortythree, he had built up the largest monopoly the world has ever seen-the great Standard Oil Company. But where was he at fifty-three? Worry had got him at fifty-three. Worry and high-tension living had already wrecked his health. At fifty-three he "looked like a mummy," says John K. Winkler, one of his biographers.

At fifty-three, Rockefeller was attacked by mystifying digestive maladies that swept away his hair, even the eyelashes and all but a faint wisp of eyebrow. "So serious was his condition," says Winkler, "that at one time John D. was compelled to exist on human milk." According to the doctors, he had alopecia, a form of baldness that often starts with sheer nerves. He looked so startling, with his stark bald dome, that he had to wear a skull-cap. Later, he had wigs made-$500 apiece-and for the rest of his life he wore these silver wigs.

Rockefeller had originally been blessed with an iron constitution. Reared on a farm, he had once had stalwart shoulders, an erect carriage, and a strong, brisk gait.

Yet at only fifty-three-when most men are at their prime- his shoulders drooped and he shambled when he walked. "When he looked in a glass," says John T. Flynn, another of his biographers, "he saw an old man. The ceaseless work, the endless worry, the streams of abuse, the sleepless nights, and the lack of exercise and rest" had exacted their toll; they had brought him to his knees. He was now the richest man in the world; yet he had to live on a diet that a pauper would have scorned. His income at the time was a million dollars a week- but two dollars a week would probably have paid for all the food he could eat. Acidulated milk and a few biscuits were all the doctors would allow him. His skin had lost its colour-it looked like old parchment drawn tight across his bones. And nothing but medical care, the best money could buy, kept him from dying at the age of fifty-three.

How did it happen? Worry. Shock. High-pressure and high-tension living. He "drove" himself literally to the edge of the grave. Even at the age of twenty-three, Rockefeller was already pursuing his goal with such grim determination that, according to those who knew him, "nothing lightened his countenance save news of a good bargain." When he made a big profit, he would do a little war dance, throw his hat on the floor and break into a jig. But if he lost money, he was ill! He once shipped $40,000 worth of grain by way of the Great Lakes. No insurance. It cost too much: $150. That night a vicious storm raged over Lake Erie. Rockefeller was so worried about losing his cargo that when his partner, George Gardner, reached the office in the morning, he found John D. Rockefeller there, pacing the floor.

"Hurry," he quavered. "Let's see if we can take out insurance now, if it isn't too late!" Gardner rushed uptown and got the insurance; but when he returned to the office, he found John D. in an even worse state of nerves. A telegram had arrived in the meantime: the cargo had landed, safe from the storm. He was sicker than ever now because they had "wasted" the $150! In fact, he was so sick about it that he had to go home and take to his bed. Think of it! At that time, his firm was doing gross business of $500,000 a year-yet he made himself so ill over $150 that he had to go to bed.

He had no time for play, no time for recreation, no time for anything except making money and teaching Sunday school. When his partner, George Gardner, purchased a second-hand yacht, with three other men, for $2,000, John D. was aghast, refused to go out in it. Gardner found him working at the office one Saturday afternoon, and pleaded: "Come on, John, let's go for a sail. It will do you good. Forget about business. Have a little fun." Rockefeller glared. "George Gardner," he warned, "you are the most extravagant man I ever knew. You are injuring your credit at the banks-and my credit too. First thing you know, you'll be wrecking our business. No, I won't go on your yacht-I don't ever want to see it!" And he stayed plugging in the office all Saturday afternoon.

The same lack of humour, the same lack of perspective, characterised John D. all through his business career. Years later he said: "I never placed my head upon the pillow at night without reminding myself that my success might be only temporary."

With millions at his command, he never put his head upon his pillow without worrying about losing his fortune. No wonder worry wrecked his health. He had no time for play or recreation, never went to the theatre, never played cards, never went to a party. As Mark Hanna said, the man was mad about money. "Sane in every other respect, but mad about money." Rockefeller had once confessed to a neighbour in Cleveland, Ohio, that he "wanted to be loved"; yet he was so cold and suspicious that few people even liked him. Morgan once balked at having to do business with him at all. "I don't like the man," he snorted. "I don't want to have any dealings with him." Rockefeller's own brother hated him so much that he removed his children's bodies from the family plot. "No one of my blood," he said, " will ever rest in land controlled by John D." Rockefeller's employees and associates lived in holy fear of him, and here is the ironic part: he was afraid of them- afraid they would talk outside the office and "give secrets away".

He had so little faith in human nature that once, when he signed a ten-year contract with an independent refiner, he made the man promise not to tell anyone, not even his wife! "Shut your mouth and run your business"-that was his motto. Then at the very peak of his prosperity, with gold flowing into his coffers like hot yellow lava pouring down the sides of Vesuvius, his private world collapsed. Books and articles denounced the robber-baron war of the Standard Oil Company!- secret rebates with railroads, the ruthless crashing of all rivals. In the oil fields of Pennsylvania, John D. Rockefeller was the most hated man on earth. He was hanged in effigy by the men he had crushed. Many of them longed to tie a rope around his withered neck and hang him to the limb of a sour-apple tree. Letters breathing fire and brimstone poured into his office -letters threatening his life.

He hired bodyguards to keep his enemies from killing him. He attempted to ignore this cyclone of hate. He had once said cynically: "You may kick me and abuse me provided you will let me have my own way." But he discovered that he was human after all. He couldn't take hate -and worry too. His health began to crack. He was puzzled and bewildered by this new enemy-illness-which attacked him from within. At first "he remained secretive about his occasional indispositions," tried to put his illness out of his mind. But insomnia, indigestion, and the loss of his hair-all physical symptoms of worry and collapse-were not to be denied. Finally, his doctors told him the shocking truth. He could take his choice: his money and his worries-or his life. They warned him he must either retire or die. He retired. But before he retired, worry, greed, fear had already wrecked his health.

When Ida Tarbell, America's most celebrated female writer of biographies, saw him, she was shocked. She wrote: "An awful age was in his face. He was the oldest man I have ever seen." Old? Why, Rockefeller was then several years younger than General MacArthur was when he recaptured the Philippines! But he was such a physical wreck that Ida Tarbell pitied him. She was working at that time on her powerful book which condemned the Standard Oil and all that it stood for; she certainly had no cause to love the man who had built up this "octopus". Yet, she said that when she saw John D. Rockefeller teaching a Sunday-school class, eagerly watching the faces of all those around him-"I had a feeling which I had not expected, and which time intensified. I was sorry for him. I know no companion so terrible as fear."

When the doctors undertook to save Rockefeller's life, they gave him three rules-three rules which he observed, to the letter, for the rest of his life. Here they are:

1. Avoid worry. Never worry about anything, under any kind of circumstances.
2. Relax, and take plenty of mild exercise in the open air.
3. Watch your diet. Always stop eating while you're still a little hungry.

John D. Rockefeller obeyed those rules; and they probably saved his life. He retired. He learned to play golf. He went in for gardening. He chatted with his neighbours. He played games. He sang songs.

But he did something else too. "During days of torture and nights of insomnia," says Winkler, "John D. had time for reflection." He began to think of other people. He stopped thinking, for once, of how much money he could get; and he began to wonder how much that money could buy in terms of human happiness.

In short, Rockefeller now began to give his millions away! Some of the time it wasn't easy. When he offered money to a church, pulpits all over the country thundered back with cries of "tainted money!" But he kept on giving. He learned of a starving little college on the shores of Lake Michigan that was being foreclosed because of its mortgage. He came to its rescue and poured millions of dollars into that college and built it into the now world-famous University of Chicago. He tried to help the Negroes. He gave money to Negro universities like Tuskegee College, where funds were needed to carry on the work of George Washington Carver. He helped to fight hookworm. When Dr. Charles W. Stiles, the hookworm authority, said: "Fifty cents' worth of medicine will cure a man of this disease which ravages the South-but who will give the fifty cents?" Rockefeller gave it. He spent millions on hookworm, stamping out the greatest scourge that has ever handicapped the South. And then he went further. He established a great international foundation-the Rockefeller Foundation-which was to fight disease and ignorance all over the world.

I speak with feeling of this work, for there is a possibility that I may owe my life to the Rockefeller Foundation. How well I remember that when I was in China in 1932, cholera was raging all over the nation. The Chinese peasants were dying like flies; yet in the midst of all this horror, we were able to go to the Rockefeller Medical College in Peking and get a vaccination to protect us from the plague. Chinese and "foreigners" alike, we were able to do that. And that was when I got my first understanding of what Rockefeller's millions were doing for the world.

Never before in history has there ever been anything even remotely like the Rockefeller Foundation. It is something unique. Rockefeller knew that all over the world there are many fine movements that men of vision start. Research is undertaken; colleges are founded; doctors struggle on to fight a disease-but only too often this high-minded work has to die for lack of funds. He decided to help these pioneers of humanity-not to "take them over", but to give them some money and help them help themselves. Today you and I can thank John D. Rockefeller for the miracles of penicillin, and for dozens of other discoveries which his money helped to finance. You can thank him for the fact that your children no longer die from spinal meningitis, a disease that used to kill four out of five. And you can thank him for part of the inroads we have made on malaria and tuberculosis, on influenza and diphtheria, and many other diseases that still plague the world.

And what about Rockefeller? When he gave his money away, did he gain peace of mind? Yes, he was contented at last. "If the public thought of him after 1900 as brooding over the attacks on the Standard Oil," said Allan Kevins, "the public was much mistaken."

Rockefeller was happy. He had changed so completely that he didn't worry at all. In fact, he refused even to lose one night's sleep when he was forced to accept the greatest defeat of his career!

That defeat came when the corporation he had built, the huge Standard Oil, was ordered to pay "the heaviest fine in history". According to the United States Government, the Standard Oil was a monopoly, in direct violation of the antitrust laws.

The battle raged for five years. The best legal brains in the land fought on interminably in what was, up to then, the longest court war in history. But Standard Oil lost.

When Judge Kenesaw Mountain Landis handed down his decision, lawyers for the defence feared that old John D. would take it very hard. But they didn't know how much he'd changed.

That night one of the lawyers got John D. on the phone. He discussed the decision as gently as he could, and then said with concern: "I hope you won't let this decision upset you, Mr. Rockefeller. I hope you'll get your night's sleep!"

And old John D.? Why, he crackled right back across the wire: "Don't worry, Mr. Johnson, I intend to get a night's sleep. And don't let it bother you either. Good night!"

That from the man who had once taken to his bed because he had lost $150! Yes, it took a long time for John D. to conquer worry. He was "dying" at fifty-three-but he lived to ninety-eight!

✦ ✦ ✦

Reading A Book On Sex Prevented My Marriage From Going On The Rocks

By B.R.W.

I hate to make this story anonymous. But it is so intimate that I could not possibly use my name. However, Dale Carnegie will vouch for the truth of this story. I first told it to him twelve years ago.

After leaving college, I got a job with a large industrial organisation, and five years later, this company sent me across the Pacific to act as one of its representatives in the Far East. A week before leaving America, I married the sweetest and most lovable woman I have ever known. But our honeymoon was a tragic disappointment for both of us-especially for her. By the time we reached Hawaii she was so disappointed, so heartbroken, that she would have returned to the States, had she not been ashamed to face her old friends and admit failure in what can be-and should be-life's most thrilling adventure.

We lived together two miserable years in the Orient. I was so unhappy that I had sometimes thought of suicide. Then one day I chanced upon a book that changed everything. I have always been a lover of books, and one night while visiting some American friends in the Far East, I was glancing over their well-stocked library when I suddenly saw a book entitled *Ideal Marriage*, by Dr. Van de Velde. The title sounded like a preachy, goody-goody document. But, out of idle curiosity, I opened it. I saw that it dealt almost entirely with the sexual side of marriage-and dealt with it frankly and without any touch of vulgarity.

If anyone had told me that I ought to read a book on sex, I would have been insulted. Read one? I felt I could write one. But my own marriage was such a bust that I condescended to look this book over, anyway. So I got up the courage to ask my host if I could borrow it. I can truthfully say that reading that book turned out to be one of the important events of my life. My wife also read it. That book turned a tragic marriage into a happy, blissful companionship. If I had a million dollars, I would buy the rights to publish that book and give free copies of it to the countless thousands of bridal couples.

I once read that Dr. John B. Watson, the distinguished psychologist, said: "Sex is admittedly the most important subject in life. It is admittedly the thing which causes the most shipwrecks in the happiness of men and women."

If Dr. Watson is correct-and I am persuaded that his statement, sweeping as it is, is almost, if not wholly, true-then why does civilisation permit millions of sexual ignoramuses to marry each year and wreck all chances for married happiness?

If we want to know what is wrong with marriage, we ought to read a book entitled *What is Wrong With Marriage'* by Dr. G. V. Hamilton and Kenneth MacGowan. Dr. Hamilton spent four years investigating what is wrong with marriage before writing that book, and he says: "It would take a very reckless psychiatrist to say that most married friction doesn't find its sources in sexual maladjustment. At any rate, the frictions which arise from other difficulties would be ignored in many, many cases if the sexual relation itself were satisfactory."

I know that statement is true. I know from tragic experience.

The book that saved my marriage from shipwreck, Dr. Van de Velde's *Ideal Marriage*, can be found in most large public libraries, or bought at any bookshop. If you want to give a little gift to some bride and groom, don't give them a carving set. Give them a copy of *Ideal Marriage*. That book will do more to increase their happiness than all the carving sets in the world.

[Note by Dale Carnegie: If you find *Ideal Marriage* too expensive, here is another book I can recommend: *A Marriage Manual*, by Drs. Hannah and Abraham Stone.]

✦ ✦ ✦

I Was Committing Slow Suicide Because I Didn't Know How To Relax

By Paul Sampson
Direct-Mail Advertising, 12815 Sycamore, Wyandotte, Michigan

UP to six months ago, I was rushing through life in high gear. I was always tense, never relaxed. I arrived home from work every night worried and exhausted from nervous fatigue Why? Because no one ever said to me: "Paul, you are killing yourself. Why don't you slow down? Why don't you relax?"

I would get up fast in the morning, eat fast, shave fast, dress fast, and drive to work as if I were afraid the steering wheel would fly out the window if I didn't have a death grip on it. I worked fast, hurried home, and at night I even tried to sleep fast.

I was in such a state that I went to see a famous nerve specialist in Detroit. He told me to relax. (By the way, he gave me the same principles for relaxation that are advocated in Chapter 24 of this book.) He told me to think of relaxing all the time-to think about it when I was working, driving, eating, and trying to go to sleep. He told me that I was committing slow suicide because I didn't know how to relax.

Ever since then I have practised relaxation. When I go to bed at night, I don't try to go to sleep until I've consciously relaxed my body and my breathing. And now I wake up in the morning rested-a big improvement, because I used to wake up in the morning tired and tense. I relax now when I eat and when I drive. To be sure, I am alert when driving, but I drive with my mind now instead of my nerves. The most important place I relax is at my work. Several times a day I stop everything and take inventory of myself to see if I am entirely relaxed. When the phone rings now, no longer do I grab it as though someone were trying to beat me to it; and when someone is talking to me, I'm as relaxed as a sleeping baby.

The result? Life is much more pleasant and enjoyable; and I'm completely free of nervous fatigue and nervous worry.

✦ ✦ ✦

A Real Miracle Happened To Me

By Mrs. John Burger
3,940 Colorado Avenue, Minneapolis, Minnesota

Worry had completely defeated me. My mind was so confused and troubled that I could see no joy in living. My nerves were so strained that I could neither sleep at night nor relax by day. My three young children were widely separated, living with relatives. My husband, having recently returned from the armed service, was in another city trying to establish a law practice. I felt all the insecurities and uncertainties of the post-war readjustment period.

I was threatening my husband's career, my children's natural endowment of a happy, normal home life, and I was also threatening my own life. My husband could find no housing, and the only solution was to build. Everything depended on my getting well. The more I realised this and the harder I would try, the greater would be my fear of failure. Then I developed a fear of planning for any responsibility. I felt that I could no longer trust myself. I felt I was a complete failure.

When all was darkest and there seemed to be no help, my mother did something for me that I shall never forget or cease being grateful for. She shocked me into fighting back. She upbraided me for giving in and for losing control of my nerves and my mind. She challenged me to get up out of bed and fight for all I had. She said I was giving in to the situation, fearing it instead of facing it, running away from life instead of living it.

So I did start fighting from that day on. That very weekend I told my parents they could go home, because I was going to take over; and I did what seemed impossible at the time. I was left alone to care for my two younger children. I slept well, I began to eat better, and my spirits began to improve. A week later when they returned to visit me again, they found me singing at my ironing. I had a sense of well-being because I had begun to fight a battle and I was winning. I shall never forget this lesson. ... If a situation seems insurmountable, face it! Start fighting! Don't give in!

From that time on I forced myself to work, and lost myself in my work. Finally I gathered my children together and joined my husband in our new home. I resolved that I would become well enough to give my lovely family a strong, happy mother. I became engrossed with plans for our home, plans for my children, plans for my husband, plans for everything-except for me. I became too busy to think of myself. And it was then that the real miracle happened.

I grew stronger and stronger and could wake up with the joy of well-being, the joy of planning for the new day ahead, the joy of living. And although days of depression did creep in occasionally after that, especially when I was tired, I would tell myself not to think or try to reason with myself on those days and gradually they became fewer and fewer and finally disappeared.

Now, a year later, I have a very happy, successful husband, a beautiful home that I can work in sixteen hours a day, and three healthy, happy children-and for myself, peace of mind!

✦ ✦ ✦

Setbacks (*)

By Ferenc Molnar

Noted Hungarian Playwright "Work is the best narcotic!"

Exactly fifty years ago my father gave me the words I have lived by ever since. He was a physician. I had just started to study law at the Budapest University. I failed one examination. I thought I could not survive the shame so I sought escape in the consolation of failure's closest friend, alcohol, always at hand: apricot brandy to be exact.

My father called on me unexpectedly. Like a good doctor, he discovered both the trouble and the bottle, in a second. I confessed why I had to escape reality.

The dear old man then and there improvised a prescription. He explained to me that there can be no real escape in alcohol or sleeping pills-or in any drug. For any sorrow there is only one medicine, better and more reliable than all the drugs in the world: work!

How right my father was! Getting used to work might be hard. Sooner or later you succeed. It has, of course, the quality of all the narcotics. It becomes habit-forming. And once the habit is formed, sooner or later, it becomes impossible to break one's self of it. I have never been able to break myself of the habit for fifty years.

✦ ✦ ✦

---

* Reprinted with permission of the author, from Words to Live By-A Little Treasury of Inspiration and Wisdom, published by Simon and Schuster, Inc., copyright, 1947, by William Nichols.

I Was So Worried I Didn't Eat A Bite Of Solid Food For Eighteen Days

By Kathryne Holcombe Farmer
Sheriff's Office, Mobile, Alabama

Three months ago, I was so worried that I didn't sleep for four days and nights; and I did not eat a bite of solid food for eighteen days. Even the smell of food made me violently sick. I cannot find words to describe the mental anguish I endured. I wonder whether hell has any worse tortures than what I went through. I felt as if I would go insane or die. I knew that I couldn't possibly continue living as I was.

The turning point of my life was the day I was given an advance copy of this book. During the last three months, I have practically lived with this book, studying every page, desperately trying to find a new way of life. The change that has occurred in my mental outlook and emotional stability is almost unbelievable. I am now able to endure the battles of each passing day. I now realise that in the past, I was being driven half mad not by today's problems but by the bitterness and anxiety over something that had happened yesterday or that I feared might happen tomorrow.

But now, when I find myself starting to worry about anything, I immediately stop and start to apply some of the principles I learned from studying this book. If I am tempted to tense up over something that must be done today, I get busy and do it immediately and get it off my mind.

When I am faced with the kind of problems that used to drive me half crazy, I now calmly set about trying to apply the three steps outlined in Chapter 2, Part One. First, I ask myself what is the worst that can possibly happen. Second, I try to accept it mentally. Third, I concentrate on the problem and see how I can improve the worst which I am already willing to accept- if I have to.

When I find myself worrying about a thing I cannot change -and do not want to accept-I stop myself short and repeat this little prayer:

"God grant me the serenity
to accept the things I cannot change,
the courage to change the things I can,
and wisdom to know the difference."

Since reading this book, I am really experiencing a new and glorious way of life. I am no longer destroying my health and happiness by anxiety. I can sleep nine hours a night now. I enjoy my food. A veil has been lifted from me. A door has been opened. I can now see and enjoy the beauty of the world which surrounds me. I thank God for life now and for the privilege of living in such a wonderful world.

May I suggest that you also read this book over: keep it by your bed: underscore the parts that apply to your problems. Study it; use it. For this is not a "reading book" in the ordinary sense; it is written as a "guidebook"-to a new way of life!

✦ ✦ ✦

NOTES

NOTES

NOTES

NOTES

NOTES

NOTES

NOTES

NOTES